AMERICAN ROAD RACE SPECIALS 1934-70

Glory Days of Homebuilt Racers

Allan Girdler

ECHO POINT BOOKS & MEDIA, LLC

Copyright © 2014 Allan Girdler

Published by Echo Point Books & Media,
www.EchoPointBooks.com

ISBN: 978-1-62654-933-3

Cover image from the 2010 Watkins Glen Vintage Grand Prix by Daniel Vaughan,
www.ConceptCarz.com

Cover design by Adrienne Nunez,
Echo Point Books & Media

Specific manufacturers, model names, and other verbiage are used for identification purposes only
and are the property of the trademark holder. This is not an official publication.

Printed in the U.S.A.

Contents

	Acknowledgments	4
	Introduction	5
1	*Por le Sport* 1934-1940	7
2	Hot Rods Versus Sports Cars 1947-1953	14
3	The Airport Era 1954-1955	38
4	Sizzling Stovebolts 1956-1958	70
5	The Professionals 1959-1964	95
6	At the Peak 1965-1966	118
7	The Special of Specials: Chaparral 2	171
8	Back in the Old Backyard 1965-1970	188
	Epilogue	206
	Index	207

Acknowledgments

One of philosophy's basic questions is, Do people make history or does history happen to people?

In the case of road racing and the homemade sports car, I have to vote for the people. If it hadn't been for certain men in certain places at just the right time, we wouldn't have had what we had.

So, acting in behalf of all of us who drove, tuned, reported or watched during the time of the special, this book is dedicated to Max Balchowsky, Miles Collier, Briggs Cunningham, Jim Hall, Ken Miles, Roger Penske and Lance Reventlow.

In a more personal way, this book would have been the poorer except for help from Jane Barrett, Pete Brock, John Burgess, Bill Greene, Jan Jones, Otis Meyer, Ak Miller, Dale Miller, Phil Remington, Bill Stroppe and Dave Ward.

Allan Girdler

Introduction

Trick question: When John R. Bond headed for California to make *Road & Track* the creative force behind the American sports car movement, what was he towing behind the family sedan?

Unlikely but true answer: His 1932 Ford roadster.

Equally tricky question: When Briggs Cunningham, dean, patriarch and pioneer of American sports car racing, decided to improve a wrecked Mercedes-Benz, what did he do?

Even less likely answer: He powered it with a souped-up Buick engine.

Obvious conclusion: In the beginning, American sports car racing and road racing were not for snobs, nor for chaps unwilling to get grease beneath their fingernails. And their machines owed as much to hard work and late nights as to pedigrees and deep pockets.

Our subject here is specials, which in racing parlance means racing cars built one at a time to meet a set of racing rules. Specials aren't mass-produced; they usually aren't offered for sale; and in the early days at least they came from backyard shops and garages, and from guys who figured they knew more than the professional builders did—which, as we'll see, they often did.

But there's more to it than that. Today's folklore holds that the sports car arrived from Europe, the province of wealthy dabblers and slick professional teams who proceeded to show the locals how the motorcar worked.

The truth is a lot more complicated (and more fun).

Begin with two strange facts. First, there was no road racing in the United States. Second, there were no sports cars.

Both statements need some qualification. There were a few road races in America back when the car was new. There were also several odd events later; for instance, the unique beach-road race held on Daytona Beach in the mid-thirties, and the races held on public roads in Elgin, Illinois, for supposedly stock roadsters in the same era. But while the Europeans were setting up public roads for private amusement, as in city-to-city races, the Mille Miglia, Le Mans and so on, our form of democracy made that impossible. The American public owned the roads, while at the same time their devotion to profit meant that promoters didn't want to have races where there was no way to sell tickets.

Thus, American racing quickly became a professional sport, usually conducted on converted horse tracks or fairgrounds ovals. There were little cars and big cars, but there was just one major event, the Indianapolis 500, and the cars that mattered had room for only one person.

At the other end of the enthusiast scale, there was a hobby that for reasons now obscured became known as Hot Rodding. The participants were generally young, male, working class and braver than lions. They stripped secondhand cars, almost always Fords, and did things to them the engineers couldn't believe. They also hit speeds the engineers couldn't believe.

The best of these men were true engineers, self-taught or not. They graduated into professional racing, especially when tough economic times, such as the Great Depression, generated rules that allowed stock-block engines into professional classes.

We in the United States aren't supposed to have social classes—and we don't, not formally anyway. We do have social and economic levels, though, and we have old money and new money and manners and so forth. For some reason, guys who liked cars and liked working on them were generally from the middle or working classes in America and in most of Europe. The car was a tool of work or a family treasure and the way you could tell how good a car was, and how important its owner thought he was, was to check on how big the vehicle was. Bigger meant better, period. Street racers, known then as Scorchers, careened around in stripped little cars. Other people worked for bigger sedans.

This was generally true in what's now the developed world, that is, the United States and continental Europe. There were occasional two-place cars and plenty of powerful engines but in general, driving wasn't a sport and cars weren't sporting.

There was one major exception: England. It did have a class system, and had relied on it for centuries. One of its features was that suffering builds character, so the higher one was, the more one indulged in habits like driving small cars with no weather protection—the smaller and less comfy the better.

In short, the English invented the sports car, a car in which getting there was all fun and miserable transportation. With the same logic, the English raced sports cars, on roads and around racetracks that mimicked roads.

Oh, there is a third strange fact here: There was no imported car market in the United States. There were countries that had automobile industries, once again America and industrialized Europe. Countries with car factories built and exported cars for and to countries that didn't. A handful of agencies in New York and Los Angeles dealt in imports in the thirties, but the annual total was in three figures and the average American car owner had never even seen an import, much less considered buying one.

Well, here we are in the thirties, in an America with a long and impressive history of racing cars (just consider how Ettore Bugatti copied Henry Miller) and racing drivers, but with no sports cars, no road racing and no imports.

We're going to have to invent our own sport.
We'll have to do it ourselves.
And we'll do it with specials.

Chapter 1

Por le Sport 1934–1940

Three brothers with a taste for speed enlist their family and friends and re-invent American road racing

How and why the brothers Collier fell in love with motor racing has been lost to us, but fall in love they did.

Barron, Jr., Sam and Miles were bright and energetic young men in 1933, and they were in the most fortunate of circumstances. The United States, indeed the rest of the world, was in the depth of a financial depression, while the Colliers' family fortune, owing surely to lack of greed in the boom years, was intact. The boys went to the right schools, had the right friends and played the right sports . . . except that from somewhere they'd heard of road racing and

Looking just barely old enough to drive, Miles Collier grins from behind the wheel of his supercharged MG. Collier Automotive Museum

sports cars. Their parents were indulgent, and in their early teens the brothers and pals raced (if that's not too strong a word) around the family's New York estate in motorized buckboards, sort of that day's equivalent to an underpowered go-kart.

They had a lot of fun, never mind that the vehicles topped out at perhaps 20 mph and the driveways and service roads weren't wide enough to allow side-by-side racing. So in 1933, when the Colliers were old enough for real cars, they founded the Automobile Racing Club of America (ARCA).

Again, one wonders why. Barron Collier, the oldest, had an MGJ2 Midget; Sam Collier had an overgrown buckboard with a Henderson motorcycle engine; and the other members of the club had English Austins. The club was a handful of young men who raced road cars on driveways, but even so they adopted the terms and tactics of international sport, complete with a club newsletter.

The other half of the equation: the flathead Ford in battle dress. One of a handful of road races in the United States, the Elgin, Illinois, event was contested (and won) by stripped Ford roadsters. As you can see, this was real racing on public roads, and as you can infer, the Fords were much faster than they had any right to be. Henry Ford Museum

From that newsletter, the *ARCA Journal*, we can glean the principles from which our sport will grow: "Road racing, recalling the days of the Vanderbilt Cup, used to have a strong hold on the American people.... It is the policy of the Journal not to deride nor underrate American racing, much less to criticize the courageous daring of the drivers. Rather, it is to encourage road racing . . . while the club is young we offer you, gentlemen, *road racing for amateurs*." The italics were the writer's, but the ideal would be a major point in all that came later.

Basically, the club combined two contrasting slogans. One, from England's Brooklands, was The right crowd and no crowding. The other, from American street racing, was Run what you brung. ARCA had chapters in Boston and New York, and the Colliers built a dirt circuit, sort of an indented oval, on family land. It was 0.7 miles around and the lap record was about 35 mph. The races were popular for their time and place, and other tracks were built, in New York and Massachusetts, at the same time.

The racing was low-key, at the least. Briggs Cunningham, an early member back in the drivewayracing days, recalled that if you slowed down as you roared past the Collier kitchen, the cook would lean out and hand you a Coke. Honest.

Now we come to the cars. Presumably to ensure a good supply, the Colliers established a company to import and even sell MGs in the United States. Several members had Bugattis, even, and there was a sprinkling of Amilcars, Rileys and Austins.

Thing was, because the racing was low-key and sporting and because the purpose of the club was to entertain the members, just about any roadworthy car was allowed to compete. On one end of the scale, two of the clan towed their racing car to the track behind a big Auburn roadster. When the entry was smaller than expected, the race organizers talked the team into running the tow car in the race. OK, the driver slid off course while hopelessly behind, but it was good sport.

At the other extreme, professional road races were held for outwardly stock cars in the thirties, notably at Elgin. A semi-pro team of Willys roadsters and at least a dozen souped-up Ford roadsters ran those races, and several of both turned up, were welcomed and even won in ARCA events. If the new Fords were faster than the old Bugattis or tiny MGs, well, nobody minded much.

This leads us to the first ancestral special.

Old Gray Mare

For its first fifty years, the sport now known as hot rodding was the province of Ford. This was partly because Ford made a quality product, using the best steel and so forth, so Fords held up under the stress of modification and racing—and partly because there were so darned many Fords that they *had* to grab the lion's share of the market.

At any rate, during the twenties and thirties there were professional classes, usually local but sometimes national, for stock-block racing cars. Some of the cars were single-seat, others conformed to rules requiring that a mechanic ride along and so the cars had to have two seats. (One wonders why there was once a rule that some poor wretch must bet himself on his driver's skill and luck!)

There was a good supply of overhead-valve, overhead-camshaft and even multiple-valve cylinder heads, along with cams, exhaust systems, carburetors and so on, for the Ford T and A engines, which began life in humble and understressed side-valve form.

And there were sports car fans who hadn't inherited money. This begins a theme we'll see again. John C. Reuter, who entered racing in ARCA with a tired Bugatti (later expensively restored into a fast Bugatti), thought there was a more practical way. He teamed up with Lem Ladd, a sharp mechanic and owner of a Boston garage, and they built their own car.

Reuter and Ladd listed nineteen different brands that had contributed to the thing, but mostly, as in suspension and engine, brakes and the like, it was Ford. The first engine was a modified Model T; then they switched to an A, and later to the V-8.

When it first appeared, the car had barely any bodywork, so it was called Fatima. According to Reuter's account in *American Road Racing*, a history of ARCA, the car then got a runabout body, mostly hood and cowling for the firewall. Later still, the car was named the Old Gray Mare. It was one of the older cars in the club by then, and had been fitted with a slick version of a sprint car body—that is, no doors and at first no fenders, but with a streamlined front and tail.

Just to keep the record straight, the car had a normal frame of channel iron, live axle front and rear, engine in front driving the rear wheels, drum brakes and wire wheels. All state of the art, which when the term was coined meant normal and average, rather than the leading edge of whatever field was involved.

The Old Gray Mare broke down in its first race, but became reliable and then got fast. This was a two-stage operation. The first came when the Model A engine was replaced with a V-8. That was a gamble at the time, because the A engine was being used in professional racing, with all manner of equipment offered, and the V-8 was still new.

The V-8 showed potential but at first it wasn't understood. The Elgin road race car mentioned earlier was reputed to have 120 bhp at the rear wheels. The owner was Joel Thorne, later famous at Indianapolis, and he had the money to buy secrets; Ladd and Reuter, and later John Burgess, didn't.

The flathead Ford V-8 had some strong points, the main one being that it was hard to break, but it had a handicap because the exhaust ports routed the hot exhaust up to the center of the engine's vee, then

Thundering out of the turn on stage left is the Old Gray Mare, probably the most successful if least photographed of the specials that raced in ARCA and early SCCA events. The car was assembled from used parts, mostly Ford, and when the tuners solved the mysteries of keeping the engine cool it set records for years. The setting here is Bridgehampton, 1952, and the car in the lead is a Formula One Maserati, pre-World War II but still competitive. Again, this was real racing on real roads, spiced here by the mixture of racing cars, sports cars and tiny Formula Three single-seaters. Road & Track

back through the water jacketing to the outside of the vee. Somebody once calculated that you could heat a three-room house with the energy and heat subtracted from the exhaust by this routing. And of course that heat and energy wasn't available to drive the car, while the radiator had more work than it could handle. The side-valve Ford V-8 was famous for overheating, both in general and in the Old Gray Mare in specific.

But thanks to patient work, and reworking of the passages and the water pump and some additions to the hoses and pipes and radiator, the engine was cooled and the car was competitive. In 1935 it set a record climbing Mount Washington, New Hampshire, and it was second in ARCA standings that season. When it did break, which happened for the next few years, the car often set lap records before steaming to a stop. The Mare set another Mount Washington record in 1938 and won Ladd the seasonal championship that year. Reworked again and again and fitted with road gear when and where required, it was raced up into the fifties and was still fast then.

Bu-Merc

Briggs Cunningham has always been the model of a sports enthusiast, the sort of man who would and did take his six-meter racing yacht to Europe on his honeymoon, in case he'd have some extra time. He went on to skipper the American defender of the America's Cup in 1958. What he did, he did seriously and he did well.

We first met him zooming around the Collier home in his MG, yet Cunningham was more interested in sailing than cars back. Since both forms of competition were held on weekends and he never put cars first, Cunningham wanted a true road car that could be raced.

There were so few imported cars in the United States at that time that when a careless driver put his SSK Mercedes-Benz upside down along a Pennsylvania highway, Cunningham had heard about it before the wheels stopped spinning. (The rumor was that the car was owned by a member of the Wannamaker family, Cunningham said years later, but that was never confirmed.)

Zumbach, the dealership in New York, had picked up the pieces, but the car was beyond normal repair. "I'd always been keen on Buicks and I wanted to build a typical hot rod," Cunningham recalled. So he paid $100, or perhaps it was $150, for the Mercedes' radiator and body and a few other bits.

Cunningham already owned a Buick, with the big inline eight. He delivered the Buick and the Mercedes parts to a shop called Banthin-Lyon, run by two men who'd worked for LeBaron and gone out on their own. Sure, they said, we can put the Mercedes body and grille on the Buick. Before that, though, they sectioned the body and grille to make them lower.

Cunningham knew Charles Chayne, the chief engineer at Buick, and he got a special cylinder head and carburetor. The engine's compression ratio was 9.5:1, radical for those days. The Bu-Merc was never formally tested, but estimates from the time had the car hitting 60 mph in seven seconds and change.

It was a fast car. It was also a true road car with all equipment. Cunningham really did use the Bu-Merc on the road during the week and when not racing—even though some of the road use was for pranks like street racing Ralph De Palma and the racing legend's hotted-up Chrysler.

The Bu-Merc was too big for the little circuits on which the club began racing, and for the genuine race courses or round-the-houses events that followed. On the other hand, many of the ARCA events were han-

The Bu-Merc, harassing a Bugatti at the 1940 World's Fair. The car probably was as big as it looks here, and about as agile, but Miles Collier managed to work it into second place until crashing into a light pole; the race was run in the parking lot. Collier Automotive Museum

dicaps, with the really fast and really racing cars required to do extra laps. This gave the little sports cars and the big odd machines, like the Bu-Merc, an excuse to at least enter.

The car's big moments came in contrasting places. The first was in 1940, when ARCA's influential members somehow persuaded the World's Fair management to allow a World's Fair Grand Prix, a run-what-you-brung race in and around the roads and parking lots of the World's Fair being held on Long Island in New York. The fair was a world event, but even so club members were permitted to enlist themselves and their friends and to race a collection of outmoded Grand Prix cars and sports cars and a few specials, while thousands watched. A real racing car, an Alfa Romeo GP car driven by Frank Griswold, won, which was fair. But Miles Collier worked the Bu-Merc as high as second before hitting a light pole, which was incredible (running second, not hitting the pole).

The car's second big moment was just as amazing. It came when Cunningham himself, in his first formal race, put the car into second place at the finish of the first Watkins Glen race, also in New York.

Ardent Alligator

Because the Colliers were kids when this began, and were romantics to the core, one can forgive some of the cuteness of those early days. Even when they extended themselves to names like the Ardent Alligator.

The car began life as a racing Riley, a two-place English car of great pedigree and little engine. It had been owned and raced by Freddy Dixon, a legendary tuner and builder, and the Colliers bought it when they got faster than their MGs could stand.

The Riley was a good little car, but in racing as in boxing, the good little guys generally get trounced by the good big guy. So the Colliers also bought a Ford, or

Up close the Bu-Merc doesn't look quite as big. The basic formula was simply to cut down a Buick and a Mercedes body, then pop the body atop the domestic chassis after using hot rod techniques on the big eight-cylinder engine. It wasn't the first such combination, nor will it be the last. That's Briggs Cunningham at the wheel. Note the license plate. Collier Automotive Museum

The Ardent Alligator, Miles Collier impatiently at the wheel, Watkins Glen, 1949. Visible here is the classic Ford flathead racing version, with finned cylinder heads and dual down-draft carburetors. Dave Ward and Bill Greene

Miles Collier shoves the car out of the hay bales, after the Alligator was a little too ardent. The Riley was a large car for its engine size, so the engine swap was easy. This is Watkins Glen, 1950. Road & Track

some say it was a Mercury, which had a slightly uprated engine, and the Riley got a flathead V-8, modified of course, and a Ford gearbox, rear axle and so on. It didn't look very different, but it was heaps faster.

By the beginning of World War II, and even more by the early postwar period, racers and hot rodders had come to understand, even love, the awkward flathead engine and to get incredible power from it. The Alligator's finest hour came in 1949, when Miles Collier won the big race at Watkins Glen.

A few years later, when he sold the car, Collier wrote a long and detailed letter to the new owner, telling him just what parts the car had, and how they'd been improved, and how to re-jet for using alcohol fuel, and on and on. Why? Because Collier, the ostensible aristocrat, wanted the old Ford-powered special to keep on beating all the exotics and the overhead-valve V-8 powered specials then coming onto the scene. (Which it did, for a short time anyway.)

ARCA's history effectively ended with the start of World War II. No events were held during 1941, in part because many of the young men in the club were already signed up with various reserve units. The war hadn't begun for the United States, but the club could see the writing on the wall. According to the newsletter, the members felt required "to bend every effort toward preserving certain standards of conduct without which neither our sport or any other may exist."

Nobody ever made a better understatement than that.

Name that talent

When the Colliers created ARCA, they were naturally amateurs and just as clearly romantics, to the extent that they used the European quirk of making up names for themselves, a tradition born in the days when a chap could lose his inheritance if caught driving in races.

When the SCCA was formed it retained the amateur clause, to the point of banning not merely prize money but drivers who had raced for a living. At the same time, it prohibited anyone who used a false name.

Well. A top driver in the United States before World War II used the fictional name Ted Tappett for personal reasons. When he began driving for Cunningham in SCCA races, he obeyed the rule and used his real name, Phil Walters—a name under which he'd never raced for money, so he wasn't banned.

Did everybody know? Heck yes, but those in power who knew didn't care, and those who would have cared weren't in power.

Back on a sporting note, we've seen that the homemade car, powered by domestic iron, had an honored place in those early days. We can go beyond that. In the ARCA magazine, Miles Collier said that when one planned a special, that special should represent "the school of thought that one cannot find too large a motor to go into too small a chassis."

Nobody ever expressed that one better, either.

Chapter 2

Hot Rods Versus Sports Cars 1947–1953

The rest of the United States discovers the sports car, the SCCA is formed and organizes races, and homemade cars win despite (or because of) their ancestry

Conventional folklore holds that the sports car movement began in America because homecoming GIs recalled the racy little cars they'd seen in England and wanted to drive them at home.

Conventional wisdom is out to lunch.

Consider first that while the Yanks were "Over There," the English had no spare fuel for fun. Next, the production of MGs et al. was counted in handfuls, so even if all the sports cars had been zipping around, they wouldn't have been enough to have been seen by the Yanks, even if said Yanks hadn't had other things to do.

Rather, the sports car movement came to the United States as an indirect result of the war. What the war did mostly was end the Depression, in the United States and in Europe. But Europeans were devastated and Americans had their industrial machine running again. There was a tremendous

War of the Worlds, circa 1950. This is Carrell Speedway, a dirt oval south of Los Angeles and the scene of much early West Coast action. They raced sports cars here because the speedway was literally the only track in town. The front car is the Baldwin Special, and what you see is what it was: a widened sprint-style body, with road gear, and with Ford suspension and V-8 engine. The car just behind it is similar. It's the Cannon Special, and it had the same suspension and engine but used a cleaner body. Road & Track

East Coast racing began in town. This is Watkins Glen, 1949, the start of the Grand Prix. In the foreground is a Ford-powered special named Merlin and next to it is a stripped 2.3 Alfa. (In the Grand Prix race, you didn't need to actually be street legal.) Neither car finished the race. Dave Ward and Bill Greene, Watkins Glen Racing Museum Archives

pent-up demand for material goods, fueled not only by time but by the notion that by golly, we'd earned it.

There was a tremendous expansion on several fronts. The Baby Boom began in 1946. The GI Bill of Rights allowed millions of veterans to go to college, where previously less than ten percent of the population could afford to go.

Cars became affordable as well and thus popular, and we all adopted the notion that driving was fun. In the South, stock car racing was invented and became a mass movement. In the Midwest, it was sprint cars and midgets. California saw the popularization of hot rods. And in the East, the imported sports car was discovered.

A special situation existed here. England was exhausted, in material as well as emotional terms. The population was tired and angry and the economy was in sad shape. The only salvation was to export, and to find new markets to replace the fading empire.

The New Englanders' wish and ability to buy and the Englanders' wish and need to sell met in the form of MGs, Jaguars, Triumphs and the like, seconded by a handful of Italian models. This first import boom was not hampered by a demand for new cars that vastly

West Coast legend 2Jr was a stripped MG that Bill Stroppe equipped with a souped-up Ford V-8 60, an economy engine that neatly fitted into SCCA class racing. Driver here is future world champion Phil Hill. Dean Batchelor

exceeded the supply, or by the lack of new models when you could find one to buy. Nor can we forget the class issue: ARCA members were sportsmen, and they weren't snobs, but at the same time they were rich more often than not, when the spectators and readers of newpapers were poor—but now, for the price of a Buick you could have a genuine MG.

At any rate, the sports car arrived on the East and West Coasts. At the same time, ARCA sort of faded away. The most active members, the Colliers, were still active, but some new people had larger ideas and a group of enthusiasts from the Boston area formed the Sports Car Club of America. The SCCA's actual formation, as in meeting and agreeing, took place during the war, with the formal part and the recruiting of members coming soon after the boys came home.

The SCCA was more formal, and more open, than ARCA. It was created to encourage competition as well as social events, and it was established with two major goals: first, there would be a national club with regional chapters, and second, racing would be for amateurs. Make that amateurs only. This was done for reasons of sportsmanship rather than snobbery, as the early SCCA events were open to such cars as the 1941 Ford convertible that beat the Bu-Merc in a hillclimb in 1947.

The West Coast had a southern California chapter of the SCCA. It also had the California Sports Car Club, along with some other groups that made it into the pages of *Road & Track* a few times and then faded away.

Perhaps the best West Coast driver of his time was Bill Stroppe, who began as a speed-crazed kid at the dry lakes. Stroppe graduated to midgets and sprint cars, first as a driver and then as a tuner and builder because he made more money that way.

Stroppe's reputation attracted a sports car enthusiast who had him build a souped-up MG TC, with driving chores to be shared by the builder and one Phil Hill, another kid with a thirst for speed. West Coast racing was being done in some odd places, such

Classes for displacement meant you could take a little Panhard sedan, as this car's builder did, and strip all the extras, like the body, and then use the front-drive engine and gearbox and brakes and such and make a neat little torpedo body and go racing. This car was never famous but it was still running some 40 years after it was built, and the owner was still having fun.

as dirt ovals and dirt tracks laid out on private land in the desert.

There were some rule conflicts, of course. In general, the SCCA followed the dictates of the Federation Internationale de l'Automobile (FIA), the international controlling body for car racing. There were displacement classes, with A being for engines displacing eight liters or more, B for engines between five and eight liters, C for engines between three and five liters, and so on. There were also requirements for sports cars to carry a passenger, have road equipment and the like. The SCCA followed these rules, with specific exceptions. Cal Club, as the rival became known, was less stringent.

Said Stroppe: "The SCCA was all foreign cars. Cal Club brought in the backyard guys. I was one of the backyard guys." And, he added, there was a constant dispute over running the homemade cars at SCCA tracks if those cars lacked, say, upholstery on the doors, or perhaps an emergency brake.

As a result—or maybe it was the cause not the effect—there was in those days a sense of animosity between the sports car and hot rod crowds. The import fans said the domestic products were cheap and crude, the home boys called the imports "tea baggers." And worse. As noted, this wasn't true when the sport was re-invented, but even so the rivalry was to have a powerful and lasting effect on road racing, and on the machines with which racing was done.

Road racing as a major sport was re-invented in 1948, at the little upstate New York town of Watkins Glen. The town catered to tourists. Its leaders were interested in sports cars, and they could lay out a course, mostly paved but with dirt stretches, that could be closed to the public for a day or so without major complaint.

The first race's winner was Frank Griswold with a supercharged Alfa *sedanette*. Griswold was the

Surely the oldest car in this new sport was the Huntoon-Brundage Ford-Duesenberg, driven at Indianapolis in 1931, then raced after World War II with flathead Ford V-8 and road equipment. Road & Track

winner of the World's Fair Grand Prix and an experienced driver, if not of professional caliber. Cunningham came second in the Bu-Merc but recalled the race as no contest: "He [Griswold] ran off and disappeared."

Satcher, Payne, Darwin and Huntoon Specials

Great minds aren't the only ones that think alike. The flathead V-8 was common and so was the two-place sprint or Indy car; the combination was an efficient, perhaps even inexpensive, way to build a special.

One such was the Satcher Special, from D. G. Satcher. This was a California car, being a Miller sprint car frame topped by the parts of three different race cars and powered by a 1937 Ford V-8 with twin carbs and dual (why isn't mentioned) magnetos.

Phil Payne had a car that was more famous, and less involved. Payne was an aggressive driver. He

Phil Payne drove this special to many best-times-of-the-day, then rubbed it in because he always made sure people knew it was a hot rod, a Mercury flathead V-8 on a 1932 Ford frame with a steel and aluminum body. Builder was Bill Baldwin. Road & Track

Proud owner Ledyard Pfund, at Watkins Glen in 1949. The frame is 1934 Ford, the engine is flathead V-8 and the handmade all-aluminum body took many hours to build. Road & Track

Edwards Special was fully enclosed and road ready, which was unusual for its day except that Edwards hoped to go into full production of his road-race car. Road & Track

invested in a really good Mercury V-8 and had a handmade body built on 1932 Ford rails, with Ford live-axle suspension and transverse leaf springs. Payne played against both sides in that he drove in sports car events but always said his car was a hot rod.

The Darwin Special—also known as the Beetle before Volkswagen made an impact or had a nickname in the United States—was a 1940 Ford engine with Ford suspension and handmade body. Some of the races were on dirt and so the classic Ford suspension was also the classic dirt car suspension.

The hot rod with the bluest blood was owned by eastern sportsman I. J. Brundage and driven by George Huntoon. The Ford-Duesenberg was the very chassis and body driven to second in the Indy 500 by Fred Frame in 1931. By 1949, it had Ford front and rear axles but with parallel springs, and a bored and stroked flathead V-8 tuned to full racing specs. It was big, and long, but had been pared to 1,900 pounds and carried doors and seats and lights to FIA requirements, and it won races against much newer cars.

But first, another outside development. During World War II, the oil and car companies had done government research and learned a lot about combustion and efficiency. They also knew that power-absorbing features like automatic transmissions and air conditioning were on the way. So they decided to improve efficiency by increasing compression ratios. That meant the faithful old side-valve engine wasn't enough. General Motors especially began work on shorter, larger, stronger overhead-valve V-8s in place of the side-valve and inline eights and sixes then in production.

GM revealed the future for the 1949 model year, in the form of overhead-valve V-8s for Cadillac and Oldsmobile. But it would be a while before the new engines filtered down. Miles Collier won the second race at Watkins Glen in the Ford-Riley Alligator, with Cunningham second in a 1948 Ferrari 166 Spyder Corsa, the first Ferrari seen in the United States. The Ferrari was a sign of other things to come, and it was a second sort of hint because it had a small engine.

The SCCA had classes for production cars and for modifieds, meaning specials or "constructed" sports cars in the SCCA's book. And it had displacement classes.

Edwards Special

The displacement classes followed FIA lines except that the SCCA fields were small, so they usually divided into big cars (over two liters, or 120 ci) and small cars (under two liters).

Sterling Edwards was a well-to-do enthusiast who knew the right people. When road racing arrived on the West Coast, Edwards signed up pioneer hot rodders like Phil Remington and master craftsmen like Emil Deidt, who did the bodies for the front-drive Blue Crown Indy cars, and they built the first Edwards.

The Edwards Special was remarkably sophisticated for its time, with independent suspension front and rear, and—Remington recalled—disc brakes years before Jaguar tried them. The discs were from Airheart, which was then making disc brakes for pure racing cars, as in sprints and such.

The engine was even more different, being a Ford V-8 60, a miniaturized flathead built by Ford during the Depression as an economy model. The Edwards engine was sleeved and de-stroked, to bring the displacement down from 135 ci to less than 120 ci, within the two-liter limit. Then came a set of cylinder heads, moving the valves from the side to overhead and creating hemispherical combustion chambers, just like the racing engines (i.e., Offenhauser) had. The heads were called Ardun in honor of their builder, Zora Arkus-Duntov, who would go on to become chief engineer for the Corvette.

Allard J-2

The rules get bent some here, because the Allard was the product of a company, and it came from England. But this can be justified because it was a small firm, mostly Sidney Allard, his sons and some helpers—and besides, they were hot rodders before there was such a word. And they loved Fords.

Sidney Allard raced and rallied Fords before World War II and built a handful of cars with flathead Ford V-8 power. So when the English roadster, as in MG and Jaguar, became popular, Allard went that route. There were two early models. The K-2 was for the road, with full fenders and comforts, and the J-2 had cycle fenders and no comfort. You could get a stock V-8, or a souped-up one, or the fullest race model with Ardun heads on the large, 240 to 300 ci version. The J-2 had a sketchy body, tiny doors and vestigial top, and it looked as mean as it was.

The Allard had one odd feature. Fords came with beam front axles and transverse springs. Allard pro-

The brave (and apprehensive?) driver is Paul Timmons, the scene is Watkins Glen and the engine is the awesome Ardun, an overhead-valve conversion devised by Zora Arkus-Duntov, who later converted the Corvette into a sports car. The hemispherical-head Ardun kept the Ford V-8 in contention when all except Ford had overhead-valve V-8s of their own. Watkins Glen Racing Museum Archives

An Allard J-2, on its way to third overall at Le Mans, 1950. Allards came with a choice of V-8s—this one had a Cadillac—and with front suspension that made them fit only for the brave or hungry. Road & Track

When Briggs Cunningham determined to enter a racing car at Le Mans, he commissioned an aerodynamicist to do the body. This is a model of the streamlined and efficient, if unorthodox, Le Monstre roadster body the engineer came up with. Collier Automotive Museum

vided semi-independent suspension by cutting the axle in half, then pivoting each half at the front crossmember. This let the outside front wheel bank into turns, which was good, but on bounce and rebound the wheels were very independent indeed and moved in arcs of their own. The sight of an Allard under full cornering load made strong men bite through their pipe stems and children cry for their mothers. The J-2 was brutal, and it won.

Le Monstre

Another kind of brute appeared in calmer fashion.

Briggs Cunningham was racing his Ferrari and enjoying it, until one day he was visited by a remarkable man, Bill Frick. Frick was a mechanical genius and a freethinker. He liked to experiment. One project was a new overhead-valve Cadillac V-8 stuffed into a neat little 1949 Ford. He called it the Fordillac, and he brought one around to show to Cunningham, who was, you'll recall, a hot rodder at heart.

Cunningham was impressed. He was also interested in France's Le Mans, the twenty-four-hour classic where he'd gone with the Colliers years earlier. And he was a patriot, witness his later defense of the America's Cup.

He put them all together: Why not enter an American car at Le Mans?

Why not indeed. Cunningham was buying Ferraris from Luigi Chinetti, himself a top driver in his day, who arranged for the Americans to be allowed to enter the tight-knit Le Mans fraternity. The Cadillac was the best engine in 1950, so Cunningham and Frick bought one sedan to be raced essentially stock, and another to become a full racing car.

This wasn't their original plan. The rules required the cars to be stock, or modified from stock, or at least based on production. Cunningham had lunch with Henry Ford II, who wasn't interested. The Cadillac folks felt the same way, but because Cunningham, Frick and party wanted the Cadillac engine, they had to use the car.

The second sedan was stripped of its body and fitted with a two-seat roadster shell, designed by an aerodynamicist from Grumman Aircraft Company. The Cadillac engine got five—yes five—carburetors as well as every racing modification available.

The name boasted its own evolution. At the onset, the car was an English pun: Le Manster—get it, Le Mans? But the French were fascinated and delighted with the car's looks and its bulk, so they modified the

In the metal, Le Monstre was large and not especially gainly, but it was slick and it was a step in the right direction. Collier Automotive Museum

Under the sleek skin was lots of work: Cunningham and Bill Frick's crew built a framework atop the stock Cadillac frame and attached the body panels to that. Collier Automotive Museum

Even the pragmatic French loved Le Monstre, which was new and different and sporting and—until the gearbox acted up—competitive. Road & Track

The Cannon Special again, in a 1950 hillclimb. Back before fiberglass, you had to build bodies in metal (either steel or aluminum), and it was much easier to be neat if you could avoid any compound curves. Road & Track

name into Le Monstre, and that's how the car was known ever after.

In the event, Sam and Miles Collier shared the sedan and Briggs Cunningham was copilot of Le Monstre with his company's general manager, Phil Walters. Both cars used three-speed manual transmissions, those being the only choice except automatic. The roadster lost all but top gear, and the sedan passed it. The sedan was tenth, the roadster eleventh. "We would have done better except that I stuffed it into a sandbank," said Cunningham—and the American return to European racing was underway.

The Cadillac engine was quickly becoming a known quantity. Allards switched from flathead Ford or Ardun-Ford to the Cadillac factory overhead-valve V-8. One J-2 with Cad power won at Watkins Glen in 1950, a race that sadly saw Sam Collier killed. A second Cad-Allard won the second race at Bridgehampton, New York, the Long Island resort town that had followed Watkins Glen into the public road racing business.

Watkins Glen was also notable in 1950 for the inclusion of a vintage class, won by a stripped Duesenberg; for the use of Le Monstre as a pace car, the roadster being too big for the circuit; and for the appearance of a special using a Volkswagen pan and engine with racing body—the first Volkswagen racer in the United States.

Altemus Auto-Banker

Creativity was running rampant. Another more radical example was called the Altemus (for the inventor) Auto-Banker, because the chassis incorporated a system of hydraulic pumps to lean the car's wheels into the corners. *Motor Trend* for July 1951 published pictures of the car with a diagram of how the system worked, and remarked that the design reduced tire scrub and might be used commercially. As one might guess, the design was never seen again.

Road racing was becoming almost too popular; there were more fans and cars and clubs than there were tracks. The East had Watkins Glen and Bridgehampton and Thompson, a small oval in Connecticut that was improved by adding a road section. The Midwest had Elkhart Lake, Wisconsin, and in California races were held on private land, as at Pebble

Not only domestic models got the treatment. This is Al Coppel's MG TC, which got lower and wider wheels and tires and a hand-hammered alloy body and ran as a modified sports car. Road & Track

Speaking of purists, this is a big old English Lagonda, into which the owner had stuffed a Chrysler Hemi V-8. In Thompson, Connecticut, in 1953, you could race what you had if you belonged to the club. Road & Track

Enthusiast Walt Hansgen owned a Jaguar coupe and wanted a racing car, so he removed the coupe body, modified the engine to semi-race specs and built this very clean roadster, with which he won major races and attracted enough attention to earn a place on the Cunningham team. Road & Track

Beach, in parks in San Diego and San Francisco, and at airports.

There were some sanctioning bothers as well. The SCCA was supposed to work with the American Automobile Association (AAA), in the form of the older club's contest board, but there were basic disagreements. The AAA banned women and liquor from the pits, the SCCA didn't. The AAA was professional, the SCCA was amateur. There may have been old-fashioned struggling for power, as well. And finally, the AAA refused to sanction races on public highways or roads.

This meant airports were the easiest, safest places to race. As it happened they provided a good way out;

This looks prosaic but it is the Altemus Auto-Banker, an SS1 (pre-Jaguar) chassis with a Studebaker V-8, an Indy two-place body and a suspension that used hydraulic pumps and pipes and linkage to lean the body into the turns. No, it didn't work; the car didn't corner any faster than conventional cars. This is Watkins Glen, 1951, and the pursuing car is Briggs Cunningham's Ferrari coupe. Watkins Glen Racing Museum Archives

Cunningham's first production sports cars were truly that. This photo was of a trip the first batch of Cunningham C-3s made from their East Coast factory to the West Coast, for testing by the magazines. The front is taped over to protect the paint. Road & Track

several serious accidents occurred, with death and injury to spectators as the cars and the will to win became more powerful than the Colliers ever imagined. The SCCA resolved to become its own organizer and subsequently became an international club, in alliance with the drag racing and stock car racing clubs.

Cunningham C-2

Cunningham had been bitten by the international bug, and by the success that his rebodied Cadil-

The Cunningham was also a racing car; witness this C-2's appearance at Pebble Beach in 1952. Road & Track

Cunningham C-2	
Wheelbase	105 in.
Length	194 in.
Tread	na
Weight	3,450 lb. (minus fuel)
Engine	Chrysler V-8
Displacement	331 ci
Claimed power	270 bhp
Top speed	152 mph (in race trim)
0-60 mph	6.3 sec.
Quarter-mile	17.1 sec. e.t.
60-0 mph	134 ft.

lac had enjoyed. He decided to take another giant step, and to construct his own cars, for racing and for sale.

Le Mans rules were a big help. Once you'd been allowed to race and if you'd finished, you could enter again, even if you'd had one brand the previous time and your own this time.

That, Cunningham had. He and Walters and the other team members laid out a beautiful full-bodied roadster. The frame was conventional, with rails running fore and aft, but front suspension was independent and rear was de Dion. The wheels were joined by a tube, but with the axles jointed and the differential mounted solidly in the frame. The new body was as lovely as the rebodied Cadillac had been odd. Power came from the new Chrysler Hemi V-8, which was larger than the Cadillac engine but which promised more power; Chrysler was involved in the project and the delivered engines had 220 bhp. The Cunningham design crew had no qualms about using parts that worked, no matter where they came from, so the C-2 used a Cadillac windshield, Oldsmobile springs and a Ford differential with two speeds aft of a Cadillac three-speed transmission.

Cunningham was one of the first to use oversized tires. Scientific theory in 1951 held that tire size didn't matter because the extra contact patch was equaled by less weight per square inch. That was wrong, but the theory wasn't disproven for another ten years.

The C-2 had everything. Perhaps too much, as the car weighed 3,400 pounds before gas, tools and tires.

Motor journalism was also in its infancy then, but the better magazines were honest and they did their own work when possible, so it's safe to reprint some of those early figures. *Motor Trend* published a test of the C-2 in January 1952. Tests then weren't quite the way they have become, but the results were reasonable under the circumstances. The quarter-mile was slower than you'd expect for the car's 0-60 mph time, but the top speed was right and the braking distance was amazing for the time, the weight, the tires and the drum brakes.

Three C-2s raced at Le Mans in 1951. The best, driven by Walters and upcoming John Fitch, worked into second overall and stayed there for six heady hours, only to slow with bearing failure and crawl home eighteenth. The other two cars retired.

It wasn't a bad showing. The cars came home and were the terror of national races for the next year, with wins at Watkins Glen and Elkhart Lake.

Barlow Simca

Imported cars had become acceptable at some levels of society, and factories from Europe—the Japanese hadn't yet gone beyond licensed reproductions—were hard at work introducing their products in the United States. One fairly obscure brand was Simca, a French staple totally unknown in American sporting circles. Roger Barlow, a dealer and driver and organizer, was determined to change that.

The Simca sedan of the day was conventional: live rear axle, inline four-cylinder engine driving the rear wheels, four forward speeds, independent front suspension and drum brakes, with a ladder frame and normal sedan body. The engine displaced 1200 cc, which made it small for even the smaller class, which had become 1500 cc.

Never mind. Barlow had the stock body removed and the frame rails widened, so the driver could sit lower and the engine and gearbox could be lower and farther back. Barlow and another man designed the torpedo body, little more than a tube with a cockpit and cycle fenders, which was built by Emil Deidt, the best in that field. The Simca got wire wheels and a mildly souped-up engine—as in increased compression ratio, dual carburetors, and polished, ported, balanced and blueprinted internals.

Barlow was outwardly casual, saying that he didn't know how much power the car had and didn't care as long as it won. It won a lot, taking the class in its first race, at Pebble Beach, and out-qualifying a good two-liter Ferrari. The Simca beat a bunch of supercharged MGs at the drag races in California and won at Carrell Speedway, a dirt oval near where the legendary Ascot oval now sits. On dirt, the Simca was faster than Jaguars and Allards, and a few weeks after that, Barlow won his class at Elkhart Lake, which was a long way from the dirt oval.

In short, it was an incredible car. It was built to SCCA rules, which by then weren't quite like FIA rules but were designed to accommodate private parties like Barlow, and it had road equipment, was even driven to at least one race. But it was never subjected to a full formal test, so we don't know the details, which is a shame.

Seifried Special

Sophistication, as in Barlow's Simca, was coming. But efficiency was still valuable. Dick Seifried, a master craftsman, began to build a track roadster, a classic racing hot rod with Ford parts; he wasn't happy with the project so he switched and built a sports car instead.

He didn't change much. The Seifried Special began with a tube frame, rails running fore and aft.

Roger Barlow's Simca was a beautiful, professionally constructed alloy body on a modified Simca frame, with a mildly souped-up Simca engine. It was expertly built and driven, and it won Simca buyers as well as races. Road & Track

The Seifried Special began life as a track roadster but was converted into a sports car, with emphasis on racing. The conversion was a neat, clean work, especially as it was a steel body, and the car won on oval tracks against rivals with more power. Road & Track

Then came a Ford beam front axle with a live rear axle from a racing company, along with sprint car steering. Engine was a side-valve Ford V-8, coupled to a three-speed Ford transmission and with the gearshift lever on the steering column (floorshifts weren't a fad in 1951).

Seifried built his own body, a torpedo with a belly pan and a minimum of compound curves. Again, the car had road equipment yet it was in no way a dual-purpose car, and wasn't formally tested.

Road & Track featured Chuck Manning's Ford Special as an illustration of how to build a good racing car at home. Manning was an engineer and analyzed all the variables before constructing this Ford V-8-powered roadster, which was able to fend off the early pure racing imports. Road & Track

But *Road & Track* reported the results of inspection at the drags, where the Seifried weighed in at 1,930 pounds and had a wheelbase of 103 in. This sounds right, as the drag race times were 14.88 seconds with a trap speed of 104 mph, results that line up nicely with a probable engine power of 225 bhp, average for a racing Ford.

The Seifried was more special than sports car. The car ran at Carrell Speedway, where it beat a choice selection of souped-up and supercharged MGs as well as the Barlow Simca and a big Allard. In fairness, the Seifried's weight distribution of forty percent front, sixty percent rear made it much more of the dirt track sprinter it was first intended to be than a road racer. Even so it was good work, and it was accepted on its own terms.

Manning Special

Homemade was taking on a new meaning. For one thing, *Road & Track* was publishing a series of articles on design, not merely showing readers how the pros did it, but helping readers—well, some readers anyway—do their own projects.

Charles (better known as Chuck) Manning was a stress analyst at Douglas Aircraft. He began doing mathematical analysis of cornering characteristics and followed that up with his own car, designed on the basis of what he'd learned from theory and example. His resulting article, published in *Road & Track* for September 1951, went beyond what most fans would be able to build but it taught everybody how much work was involved.

Probably more useful, he showed us his example.

Like Seifried, Manning used the old reliable flathead Ford V-8, equipped for racing. The car had Ford suspension and live axles at both ends, with the engine in the middle of the chassis and the driver perched just in front of the rear axle, as low and rearward as could be. Manning had still another torpedo body, again because that was the least expensive and easiest and most practical way to go.

The major improvement here was that Manning had done his homework. He'd analyzed all the basic drive configurations (i.e., front engine, rear drive; rear engine, rear drive; front engine, front drive; even rear engine, front drive), and had worked out the good and bad points for each. (For the record, there were no good points for using rear engine, front drive.)

Parkinson Jaguar

A middle ground was taken by Don Parkinson, who in effect did to an XK120 Jaguar what Cunningham had done to the Cadillac sedan—that is, stripped it, modified it, and installed a lighter and more streamlined body.

Thing was, the Jaguar was a far superior starting point. By 1951, the Jaguar factory was in the speed equipment business, with big carburetors, special cams and heads, and the like.

Nobody said you had to begin with a Ford, or even an imported sedan. Don Parkinson converted his Jaguar XK120 into this torpedo-body racing car. Cycle fenders were legal in SCCA racing for years after they'd been banned in Europe. Road & Track

By the end of 1951 there was a rough sort of parity among the various specials and the best production cars, assuming we can consider the Allards as such. Cunninghams won Watkins Glen and Elkhart Lake, Allards took Bridgehampton and Reno, Nevada, and Parkinson was the winner at Palm Springs—after Seifried's car had the pole, with Manning in the hunt but trailing Barlow's Simca, which in turn was behind Bill Stroppe in a purely hot-rodded MG with Ford V-8 60 power.

Just to keep things straight, this little jewel is a Crosley Hotshot. When sports cars arrived, virtually the only little cars sold in the United States were tiny Crosley sedans and wagons. Ahead of their time, people said. But not sporting. So Crosley introduced this roadster, with detachable doors, in a game attempt to lure sports. It was a good car, but not good enough . . .

. . . So enthusiasts Phil Stiles and George Schrafft talked Crosley out of an engine and chassis, which they modified and fitted with a tiny sprint-style body. Then they talked Le Mans into accepting their entry. They didn't win, but that had never been the point. Road & Track

Le Biplace Torpedo

This is almost too romantic to be true, but it's fact: The first Sebring race, held as a memorial to Sam Collier, was won on index of performance by a nearly stock Crosley Hotshot. (Crosley was an early subcompact car—too early as it happened—and the Hotshot was Crosley's roadster, a miniature sports car.)

Two fans, Phil Stiles and George Schrafft, were there and were impressed. It was just talk until Mrs. Stiles said, why don't you guys ask Crosley to build you a car for Le Mans? The idea was so perfectly foolish that they wrote two letters: one to manufacturer Powel Crosley, Jr., saying the factory had been invited to the race; the second to the Le Mans club, asking for an entry.

It worked. Crosley provided a chassis and engine; the club accepted the entry. The engine was given normal racing tune, with at least 45 bhp from 45 ci, and the car got a body by Floyd Dryer, a legend from his days building leaning racing motorcycle sidecars and as a bodyman for Indy cars. The Crosley got a body that looked like a scaled-down Indy Duesenberg, with two seats of course, but with clamshell fenders and a midget race car nose.

The Crosley was broken in on the highways between Indianapolis and Le Mans, squeaked through tech inspection while winning French hearts, and was running well in the race until the last-minute generator seized solid and took out the electrics and the water pump. It was a lifetime's adventure for two men, and a lesson in how much the small team can accomplish.

Meyer Special

The two-place track car—improved, of course—was still a practical proposition. Witness the car built by John Meyer. It was an older sprint car with fenders and lights, never forgotten by the author since the day I was on my way to an early race and was passed by Meyer himself, driving to the track with the toolbox bouncing in the passenger seat.

By early 1952 the airport was replacing the resort town as the venue of possibility for "road" racing. The law of averages was catching up as the average speeds of the races went up.

The new overhead-valve V-8 was also becoming standard equipment, and Fred Wacker, one of the better Allard drivers, came in for some heat because his Cadillac engine was backed by an *automatic* transmission! By this time purists were sneaking into the tent and claiming that because highly tuned little engines needed gearboxes with lots of speeds, said gearboxes were superior to types that were less work. Wacker said his Hydra-matic worked, fitted in with the lazy torque of his big V-8 and was easier to drive, so why worry?

Cunningham C4R

Inspired and instructed by his earlier efforts, Briggs Cunningham returned to Le Mans in 1952 with

Another classic, the Meyer Special. Not everybody could build or buy a handmade body. But an adept person could do what John Meyer did: use an older, two-place sprint car; install a V-8 (in this case Cadillac); and have a car that could win races and then be driven home. Road & Track

Cunningham wanted Le Mans and it required a new, more race-oriented car, so that's what he built in 1952. The C4R looks lighter, faster and meaner than the earlier model, and it was. This is Phil Walters, the former Ted Tappett, winning the main event at Floyd Bennett Field, 1953. Road & Track

The Cunningham C4R used a ladder frame and a huge Chrysler Hemi V-8. Chrysler helped the team when executives realized how many Americans cared about racing, and that they'd learn something themselves from what Cunningham was doing. Road & Track

The C4R was built in roadster and coupe form, the coupe having a K designation for Kamm, the man who discovered that a chopped-off tail was good for streamlining. The paint scheme, white with blue stripes, was the official color combination for the United States in international racing. Road & Track

the same cars, but lighter, more powerful and more graceful. There were two roadsters and a coupe, named the C4RK, the K designating a short tail in the aerodynamic style recommended by Dr. Wunibald Kamm.

Phil Walters drove the coupe and thrilled the crowd, especially those from the United States, by leading the race for the first lap. Not to be, alas. The coupe and one roadster retired while the other open car, driven by Cunningham himself for 19½ hours and gifted amateur Bill Spear for the rest of the time, was fourth overall and class winner.

The team returned to America, again, and won national races, again. It wasn't as easy this time, as a C-Type Jaguar took Watkins Glen away from the locals. Against that, Allards still ruled the West Coast, as "the potent 4-liter Ferraris [had] yet to prove their superiority," according to *Road & Track*.

Kurtis 500S

During his time, Frank Kurtis was the king of Indianapolis. Kurtis was a big man, in body and in thought. He began building racing cars when he was a kid—was underage, in fact, but could bluff his way into the pits because he was big. He wasn't afraid to experiment or to take chances, and by the early fifties if you wanted to win in American professional racing—that is, midgets, sprints or Indy—you had to have a Kurtis chassis.

When the sports car movement arrived, Kurtis did some work on a road car, a full-bodied and slightly

Kurtis 500S	
Wheelbase	100 in.
Length	na
Tread	58 in. front, 56 in. rear
Weight	2,265 lb.
Engine	Hudson inline six
Displacement	308 ci
Claimed power	160 bhp
Top speed	102 mph
0-60 mph	7.7 sec.
Quarter-mile	16.0 sec. e.t.
60-0 mph	na

The C4RK coupe was easier on the driver, but even so it was big and mean and not especially slick, chopped short tail or not. Road & Track

Cunningham declared his cars to be production sports cars, and they were made in street form as well as for racing. This C4 roadster, shown in West Palm Beach, Florida, where the cars were made, still has an air of raw power and speed. Road & Track

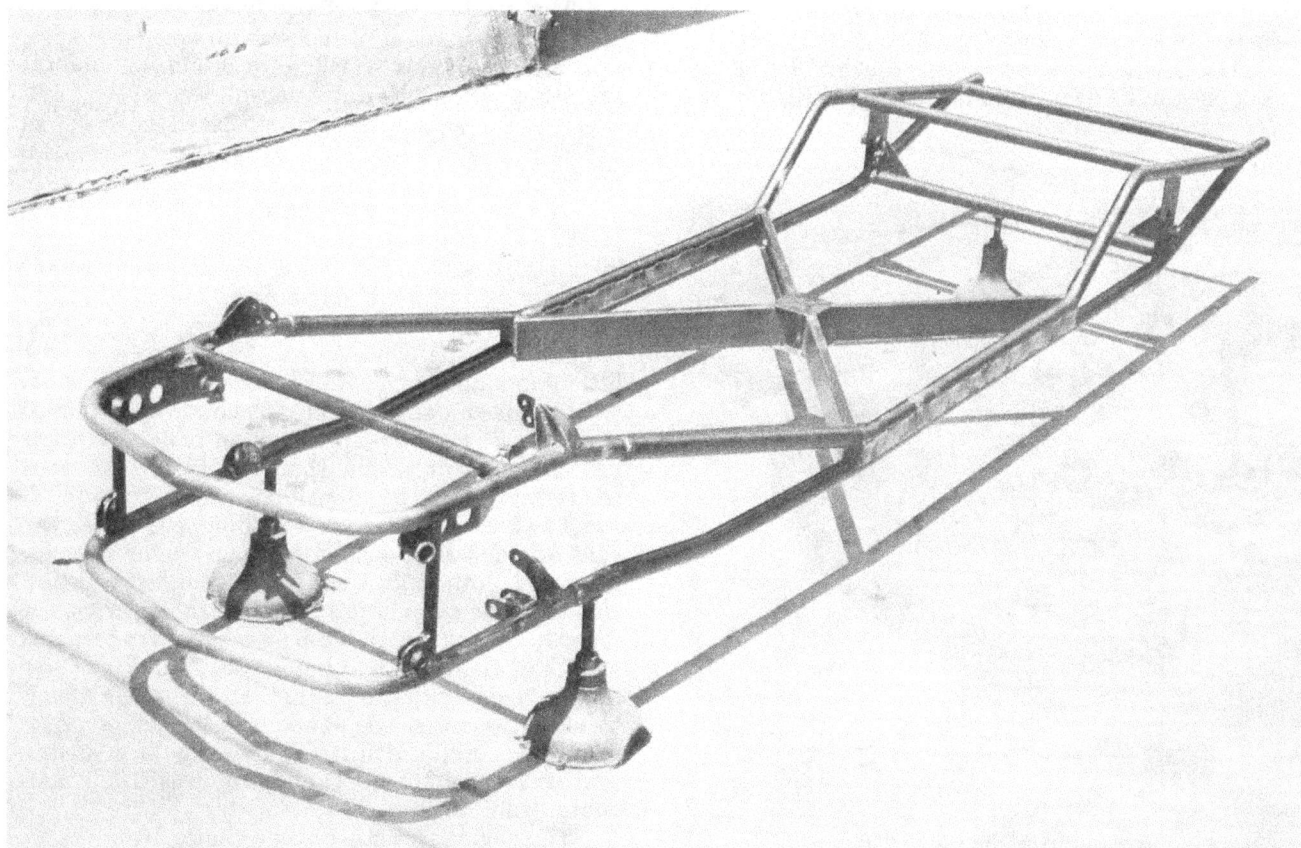

The Kurtis sports car began with this ladder frame, basically as used for Kurtis' Indy cars. Road & Track

Production Kurtises came with choice of engines. The sketchy body looked like the Indy car but wider, and used knock-off wheels. Road & Track

overdone convertible, then sold the rights to Earl "Mad Man" Muntz, the inventor of the car radio. Muntz didn't know how to build or sell cars, and the project died.

But a friend lent Kurtis a sports car, an Allard. Kurtis took the Allard for a spin, and the Allard took Kurtis for a spin. "I knew I wasn't a great driver," he said later, "but I also knew I wasn't that bad. I liked the basic idea but I decided to make a lightweight roadster that would handle."

The cockpit of the Kurtis was sparsely furnished but had full instruments, in keeping with the theme of sports racer. Road & Track

Kurtis was a practical man who knew theory but insisted on results. He was building rear-engine Indy cars before anybody else, in the United States or abroad, dared to suggest that back motors were the wave of the future.

The production Kurtis Indy car was called the 500K. It was a ladder frame with torsion bar suspension and live axles. The idea was old, but for the smooth tracks of the day a carefully controlled axle was better than a passenger-car-based independent design, so that was what Kurtis used for Indy, and for his sports car.

The first sports car version was called the 500KK. It was a kit, a rolling chassis in which the buyer could install the engine of his choice, and atop which could go any body one had laying around the shop.

There weren't many such buyers, so Kurtis expanded into the 500S, a complete sports car, with little doors and such for the rules, and widened for two side-by-side seats but in essence still a production Indy car. The sparse and Spartan body was dominated by a row of chromed bars that looked like teeth, the better to bite the opposition with.

The frame of the 500S was double channel, two rails on each side, with gussets and tubular crossmembers. Both axles were live, suspended with torsion bars, the axles located by twin trailing arms. The torpedo body was aluminum panel, with fenders that were half clamshell and half cycle. The cockpit had lowered sides with tiny doors that were there mostly to meet the rules. The simple suspension could be tuned to match the surface, allowing for a surface that was flat and smooth, as Indy was. The Kurtis contained no secrets.

The engine was still buyer's choice, with kits for Cadillac, Chrysler, Ford flathead and even Hudson, which at that time made a powerful inline six that

The most successful of the early Kurtis sports cars was this one, built for Bill Stroppe and with a shorter-than-stock wheelbase. Here Stroppe is at the wheel and in the lead. Road & Track

won stock car races even though it was a side-valve design. The kits consisted of frame, suspension, body, radiator and steering. In addition to the engine, the customer was expected to supply the transmission, rear axle, brakes, wheels, tires, instruments, paint and upholstery.

The kits would have been the best bargain, as list price was $3,000, if the buyer had an engine already in the backyard. As it happened, most of the handful of Kurtises sold were complete cars built to the customers' orders.

That was good for the buyer, but it muddies the water here because the only genuine test performed on a Kurtis 500S was done by *Road & Track*, using the first 500S sold. That car went to a Hudson dealer and was powered, as you'd expect, by a Hudson inline six. The engine had the twin carburetors and high-compression head used by the surprisingly successful Hudsons raced on the stock car circuits, but it only produced 160 bhp and the test Kurtis' curb weight was 2,265 pounds.

The *Road & Track* test results need some explanation. As the test said, the engine was used because the owner was a Hudson dealer and the car wasn't intended to be raced. The performance was quick for its time, as the big six had locomotivelike torque, but it ran out of breath about the time the car's non-aerodynamic shape became the other limit on performance. So we must note that both the Hudson stock car racers and the racing Kurtis 500s were much faster on the top end: 122 mph and 128 mph, respectively, according to records of the day.

The magazine first commented that the Kurtis' handling and steering were impeccable, as you'd expect from a racing chassis, and then commented that there was no comfort, nor were sports cars of this type expected to have any comfort. All part of the sport, or so we told ourselves then.

The real Kurtis story was elsewhere. We last saw Bill Stroppe building hot rods for the sports car races. He and his partner, the legendary mechanic-builder Clay Smith, were doing well by 1952 and Stroppe decided he'd like to get behind the wheel again. He knew Kurtis and Kurtis wanted Stroppe to have one of the first sports models, but the base price was a lot for a street racer. Kurtis donated the chassis and body, while Stroppe provided the running gear. He picked a Ford Model A rear end and a Ford flathead V-8, not because they were the best—they were woefully obsolete by 1952—but because they were what Stroppe knew, and what he had.

Stroppe gave the engine the full treatment. In still another example of great minds thinking alike, Stroppe in California performed the same sort of work with the water jackets and pump that John Burgess had done back in New England, and with the same results: 200+ hp from a twenty-year-old design.

There was one other difference. "The car looked too long to be a good sports car," Stroppe said years later. "I told Frank I wanted mine to be one foot shorter. He said no, they'd need the longer wheelbase on the faster tracks." They compromised. According to the brochure and the magazines, the 500S came with a choice of wheelbase: 94 or 100 in. But Stroppe's example had an 86 in. wheelbase.

And it worked. Stroppe and Smith were the sort of guys who'd take a flathead V-8, in a 1948 Mercury coupe, and make it run against the world's best touring cars, which they did in the Mexican Road Race. They built well and they built carefully.

Stroppe also was a talented driver, able to distinguish between fast and too fast. His Kurtis quickly became the winner, or the car to beat. It wasn't sophisticated, but it was predictable in the turns, while the strong low-end power of the Ford V-8 shot it out of the slow turns like a cannonball.

Stroppe and Smith had two secrets. One was that the Kurtis could be tuned to the track, with suspension settings and with gearing: the Ford rear end came with a wide range of racing ratios. The other was the driver, Indy star and natural Troy Ruttman.

In his sprint car days, Ruttman drove for Stroppe. One day he asked for an oversized hubcap for the right front wheel. Sure, said Stroppe, but why? Because Ruttman was able to position the car exactly if he could just tick the wall of the banked oval they were racing on. He got the hubcap, ticked the wall every lap and won.

In his spare time, Ruttman did the same with the sports car. Stroppe recalled that the Kurtis could out-corner the more powerful but heavier Cunninghams, while the equally light and agile Ferraris "had to go through the gearbox to get the power."

Caballo de Hierro

Ak Miller was a hot rodder, owner of a garage that built and maintained hot street equipment. He was also a man with an open mind, a thirst for the new and different. He held some records from the dry lakes, and he loved cruising the open road deep into three figures. And he was personable and persuasive.

The time came when the Mexican government completed paving the nation's biggest and best highway, linking the north and south borders. The government was so proud that it decided to celebrate with a race, from one end of the new highway to the other. That first year, 1950, the race was supposed to

Stroppe's engine (partially visible here) was the classic flathead Ford V-8. The Kurtis won because it worked on smooth surfaces and because the power came straight from the engine and didn't need to run through a lot of gears.

be for stock sedans, but there was so much interest and confusion and enthusiasm that it was opened to modified production cars, then sports cars.

Miller and partner Doug Harrison drove the 1952 race, properly called the Carrera Panamericana, in a stock Oldsmobile sedan. They were forced out by a failed bushing, a stupid little failure of a minor part, and they sat by the side of the road in the blazing sun for several hours.

As they sat, they swore. They swore they'd build their own car, one that wouldn't fail.

The car was named Caballo de Hierro, Spanish for Horse of Iron, and it was to be a hot rod. Make that Hot Rod. One of Miller's many pals was the editor of *Hot Rod* magazine and a founder of the fledgling National Hot Rod Association (NHRA). This was back when hot rodding was semi-criminal, frowned upon by polite society. The creators of the NHRA and staff of *Hot Rod* wanted to change that.

Plus, there was still the hot rod versus sports car feud.

Miller took the best from his familiar world, in the form of an Oldsmobile V-8. He'd already modified many examples of the make and he knew exactly how much stress the engine would take and for how long.

The frame was from a 1950 Ford, shortened to a 100 in. wheelbase but retaining the independent front suspension. The front end got stiffer springs and some

The Caballo de Hierro had modern suspension and brakes and engine, but it also had a huge and awkward frontal area, which limited top speed and buffeted the driver and passenger. This is one of the race's daily starts: from the bugs in the grille and the look on Miller's face, it's late in the race. Road & Track

Ak Miller, in the Hot Rod-*sponsored 1950 Ford frame with Oldsmobile V-8 engine and 1927 Ford roadster body, wowed the crowd in the Mexican road race.* Road & Track

The Ford-Oldsmobile really was a hot rod and it kept the track roadster tradition of finding space for every sponsor. The rules said you had to have a spare tire and there was no trunk, so the team bolted the wheel to the trunk lid. Road & Track

Hot rod versus hot rod? This is Willow Springs, 1954. A Kurtis is leading Ak Miller's Ford-Oldsmobile and another Kurtis is chasing them. The Kurtis worked in sprint races but the heavier endurance racer didn't. Road & Track

helper springing in the form of air bags. The Ford rear end was located with a transverse leaf spring, like the rear ends on traditional hot rods of Miller's experience. Steering came from a 1940 Ford. The body was classic—1927 Model T Ford roadster—and Miller swears it came from a friend's backyard, gratis.

Miller knew the Olds gearbox wouldn't hold up, so he fitted a three-speed manual from a 1937 Cadillac, backed up by an overdrive from a 1937 Nash. One thing the pioneer street and lakes racers knew was what worked and what didn't—and to heck with brand names.

The brakes were 13 in. drums from a 1935 Chrysler, adapted to Ford spindles. The wheels were Lincoln for the practical reason that Miller knew Stroppe and Stroppe was running the Lincoln team, which had tires to spare while Miller didn't—and anyway, Ford and Lincoln were part of the same corporation and all the domestic sedan racers were on the hot rod side.

The hybrid got a plexiglass windshield that was nearly straight up-and-down and a 40 gallon fuel tank. Where the Model T body didn't cover, Miller hammered and trimmed pieces from, for example, a 1937 Plymouth.

This all sounds more crude than it was. The car weighed perhaps 2,200 pounds at the curb, 2,900 topped off with fuel, water, oil and driver. The 1953 engine probably had 200 bhp, the second version cranked out 285 bhp at a relatively relaxed 4800 rpm.

Hot Rod gave the Caballo full attention. Through its pages we learned that we could put our money where our hearts were by sending one dollar to the NHRA, for Miller and Harrison and the team. (As I told Miller ten years later, when he worked for Ford and I was a newspaper reporter on the automobile beat, I never spent a dollar more proudly.)

Donors and loaners and such help aside, Miller reckoned later he'd spent $1,300 in cash to build and prepare the car. As part of the budget effort, Miller and Harrison drove the car to the 1953 Carrera Panamericana. They learned that they were in trouble. The overdrive let the engine run slowly in the top gears and that meant a low—numerically high—final drive ratio for the mountains. But running the engine slowly and the pinion gear quickly fed more heat into the pinion gear's bearings than they could stand. The differential began to howl before they even got to the starting line.

There was no way to fix it. Instead, they replaced the differential every day, during the allotted time for repairs. While Miller and Harrison were driving, the crew was scouring central Mexico for more Ford gears. Incredibly, the car finished eighth; when it was safe to run hard, the hot rod was a match for all but the very best in the race.

Caballo de Hierro ran in the Mexican Road Race again in 1954. This time it had a new Lincoln rear axle, geared right and lubricated with Clay Smith's suggestion of a mix of castor bean oil and white lead. The engine was larger and stronger and the team had money for parts and uniforms.

Caballo de Hierro was the third car in at the end of the race's first day, and the crowd went wild—never mind that the cars started at intervals so the hot rod was really fourth in the sports car class. It was homely and homemade and a clear underdog, and the fans, like fans everywhere, loved the notion of the little guys handing it to the factory teams.

Which Miller and Harrison very nearly did. The biggest Ferrari was clearly faster but the three-liter entry from Big Red wasn't able to pass at will, the way the driver had predicted before the start, while other pure sports cars, C-Type Jaguars for instance, couldn't keep up with the hot rod.

Even so, there were some silly problems. The spare tire's pressure hadn't been checked and when that tire was used, the car was all over the road until the flaw could be fixed.

There were also clever solutions. The race cars had to use Mexican gas, as famous for poor quality control then as now. The Europeans weren't prepared for this, but Miller was. He rigged a manual control for spark advance, and a mirror that let him watch the car's exhaust at speed. If the pipes got hot or began to smoke, Miller was ready to read the signs and adjust the timing.

Caballo de Hierro was fifth in the big sports car class, seventh overall at the finish. From one view the car was the worst possible for a high-speed race because it was so heavy and so non-aerodynamic: true top speed was about 145 mph, while a car with a streamlined envelope body would have hit 170 to 180 mph with the same power.

From another viewpoint, the hot rod was the best possible car for the event. It finished, and it showed the hot rodders and the sports car set, the Yanks and the furriners, what native ingenuity could do. From that day on, both sides took the other side a bit more seriously.

The end of the story is that Miller sold the car, minus engine, to a man who didn't pay what he'd promised. Then the car turned up with Ford inline-six power, of all travesties, and finally it disappeared, never to be seen again. What a pity, and what a loss.

Chapter 3

The Airport Era 1954–1955

The sport becomes a business, technology threatens mere skill and the proceeds, as they say, go to charity

As always, the Greeks had a word for it and the word was *serendipity*: the happy accident that allows two plus two to equal five.

We move back in time a year or two here. Ak Miller's Caballo de Hierro Oldsmobile-Ford special was the peak of the curve, the best showing of a pure hot rod in the last pure point-to-point race on public roads in the Americas.

The signs were up a few years before that, as there were incidents and injuries and even deaths from road racing. The crowds were getting larger and the cars faster, and the other folks in the various

It's time for the 1953 Savannah Grand Prix, and you can see both the size of the track and the size of the crowd; back then they weren't fussy about how grand the prix had to be.

In the front row are Sherwood Johnson in the Cunningham C4R and Bill Spear and Jim Kimberly in Ferraris. Road & Track

resort towns were less and less willing to have their roads closed and their lawns wrecked by swarming thousands of strangers (one cannot blame them).

At the same time, visionaries had faith enough to invest in private land and begin planning closed racetracks, with winding circuits and hills and so forth, imitation roads for racing only. Such projects take time.

The happy accident was an Air Force general named Curtis LeMay. He could have served as the prototype for a man's man, a gruff and bluff gnarly galoot right out of a western movie, except that he was also smart and brave and trained in the art of public relations. Make that politics.

LeMay was introduced to sports cars because he was a pilot and thus a natural motorhead. He commanded the Strategic Air Command (SAC), the huge bombers that provided the big stick to go with the free world's soft voice. Remember, this was in the era of the Berlin Wall and Korea—we aren't talking *glasnost* here—when the cold war could have gone hot.

This was the perfect partnership for the time and place. The SCCA had a booming sports car market and a membership growing right along with the sales curve. The sport of road racing was finally escaping from the old clubby atmosphere and racing in any form was moving away from the horse track and the fairgrounds. And then New York's Watkins Glen and Wisconsin's Elkhart Lake made racing on public roads illegal.

The SAC had bases in all sections of the United States. The bases had miles of runway, flat and smooth and long, lying there all day long and used for only a few minutes every twenty-four hours. There was a fine supply of workers—volunteers, or so the Air Force said. And military air bases came equipped with fencing and gates and so forth; one obvious problem back in the days of racing through towns was the lack of gates and fences and thus a handicap when it came to collecting tickets.

All it took was an agreement between the SAC and the SCCA. The contract was signed in 1952, and air bases quickly became *the* venue for sports car racing, albeit not exactly road racing in the Collier brothers style.

That didn't matter. Sports cars were increasingly popular and so was racing. The SCCA races at MacDill Air Force Base (AFB), near Tampa, Florida, drew 90,000 paying spectators. The Cunningham team won, the crowd was thrilled and the designated charity, a fund for enlisted men and their families, was greatly enriched.

Ken Miles' R-1 was immaculately prepared. The car wasn't styled so much as created in the image of an MG stripped for racing, while the suspension and frame were much advanced and lighter and the engine was built on a special block from the factory's secret shops. Notice the frame hoops that carried the body panels, atop the firewall, and the odd-but-effective air box for the two carburetors. Road & Track

There were some variations, as in the race at Sebring, Florida, which was sanctioned by the FIA. Sebring was a former military airport, but it wasn't part of the SAC. The only other FIA event in the United States was Indy, and that was a courtesy. Some races were also still run on semi-public land—for instance Torrey Pines, California, where late in 1952 Phil Hill won with the Jaguar XKC, followed by the Parkinson (Jaguar) Special.

Miles R-1

Another reason military airports became the place to race was simply that finding a full field was as easy as mailing the entries. The SCCA had grown in size and in expertise, and there had been some relaxation in the rules regulating who was allowed to drive and who was a professional. The SCCA definitions of production and modified (specials) weren't quite in line with those of the rest of the world, mostly because the American club wanted to encourage racing from all walks of life and depths of pocket.

So it followed that thousands of enthusiasts signed up for drivers schools and for SCCA licenses. This created a market for pure racing cars. In the smaller class, 1500 cc or less, the OSCA (a real racing car) and the Porsche (a highly tuned road car) had eclipsed the Barlow Simca. The Italian and German brands were well designed, built and driven, and they earned a lot of ink.

Time for a change, in the person of Ken Miles. He was an Englishman, a blade-faced former tank mechanic who'd abandoned his family's import-export business for the lure of racing, and who'd left home after his war exploits to make his way in the United States. Miles was an extraordinary man. Fiercely competitive and at the same time loyal, he'd brag when that would work, and hide his light under the proverbial bushel when that was the better technique.

MG was the popular brand then, being the least expensive of the true sports cars, but with its mass-produced pushrod engine it was no match for the overhead-cam, limited-production engines from OSCA and Porsche.

Miles was service manager for Gough Industries, the MG distributor for southern California. He'd done some racing and building back in England, and MG's image needed some improvement on the racing circuits, so he and his boss agreed to give new meaning to the term company car.

Miles did the designing and the construction, with help from company mechanics and employees and using the firm's premises. The car was his, in fact and in title, while the boss supplied parts and insisted that there be as many MG parts in the special as was practical. The boss also rode herd on Miles' tendency to overdo the details.

At Moffet Naval Air Station, Ken Miles gets a drink while the crew fettles and fuels. The odd tank behind Miles' head is a gravity-powered fuel canister. Road & Track

The result, called the R-1 but later affectionately known as Number 50, was much more modern than it looked. Miles began with a ladder frame of mild steel. Front suspension and steering came from a Morris Minor with fabricated arms patterned on the MGTD's and with MGTC hubs. The rear axle was TC and so were the brakes with the addition of aluminum-finned brake drums.

The work began at a lucky time, as the MG factory was busy with record attempts on the Bonneville Salt Flats in Utah and Miles was given a works engine block from that project. It came with a larger bore—72 mm compared with the 66.5 mm of the stock 1250 cc MG engine—than the production block could have survived. There was also a special crankshaft, and the MG factory had long offered racing parts, as in cams, cylinder head design, and blueprints for exhaust systems and the like. MG began life as a builder of purebred sports cars, then became an outlet for less expensive cars, in greater numbers but with sporting pretensions. The factory was always in the racing business one way or another and there was a good supply of competition parts. This made it easy for Miles and scores of other builders to use MG parts.

It also led to some confusion. In 1989 the R-1 was being raced, and was winning, in the vintage class. The engine in the car when it had been sold thirty years before conformed to the above. But in his excellent history of the car, published in *Road & Track* for February 1977, James T. Crow wrote that pit gossip from 1953 was that Miles had an even more modified engine, with 100 bhp instead of the 85 bhp with which the Bonneville-based powerplant was credited. There may have been two engines, in other words.

That would have been something Miles would have enjoyed.

Meanwhile, the special's body was classic homegrown, with aluminum sheet wrapped around the essential parts. The torpedo body had a minimum of compound curvature, and trouble and work were also spared by doing such things as tucking the spare tire (which the rules mandated) into what otherwise would have been a hole in the tail. Fenders were cycle and the space in front of the Morris radiator was filled with cut-down Morris grille.

Now comes the craft. Miles was a master of psychology, in the form of making the other guy think what you want him to think.

The R-1 was never officially weighed in its original form. The official guess in 1953, from Miles, was 1,250 pounds, while later owners figured it was more like 1,175 pounds. Miles would have liked the opposition to think his car was heavier and more powerful than it was, just so the performance would seem even more formidable.

Formidable is too weak a word. The R-1 first appeared at Pebble Beach in 1953, and won its class going away. The car ran in ten major races with Miles at the wheel, and won eight of them. The other two races ended with mechanical failure—the clutch went once and the flywheel mounting bolts sheared in the second retirement—so it's fair to say that the R-1 was never beaten by another car. The crowd loved this.

Miles had several ploys. He'd run up close behind the leader in class and harry the other driver into a mistake, at which point he'd nip past. Next week he might simply motor into the distance from the flag. Or he'd pass and repass, rubbing it in. He was a good driver, arguably as good in a small car as the sport had seen up to then.

There were emotional ties as well. First, by no accident, the R-1 looked, if not like an MG, then at least like a car built by stripping an MG of all the clumsy bits. Second, Miles had a biting wit and a reluctance to suffer fools at all, which made him a favorite of the press, which was always in need of some quotes to go along with the picture of the winner and the trophy girl. Miles seldom disappointed.

Crow said there was another factor. MGs didn't win many races in those days—at least until the production class break was, er, rearranged to take the faster cars away from the MGs—but MG was the choice of the fans, those who drove to the races and paid the tickets. They naturally enjoyed watching this brilliant collection of MG parts. Furthermore, this was 1953, and the sports car fan had likely been in World War II. He hadn't forgotten who invaded whom and who ran the camps and so on, and the sight of the German Porsche and the Italian OSCA getting whipped by the English MG let him remember who'd won the war.

The R-1 was, in the long run and the short run, a stunning success. There's no question that it didn't just win, but also sold a bunch of MGs and inspired affection in the hearts of the folks who already owned one. And if the car wasn't exactly the backyard, homespun hero Miles liked to portray, well, there

The tricky Allard J-2 was tamed some by extending the frame and moving the engine forward, making it more difficult for the tail to wag the dog. Road & Track

41

Allard next introduced the J-2X Le Mans, an informal title for the streamlined body the factory used in the twenty-four-hour race. It had the same frame as the J-2, though, and the same demanding chassis. Road & Track

were no rules against it and Miles *did* build the car he drove.

Allard JR

The arrival of innovative and expensive sports-racing cars signaled a leap in technology, and a threat to the older, simpler style.

One result was the Allard JR. When last we saw the marque, the big J-2, with Cadillac engine bellowing and front wheels waving independently in the air, was the baddest of the beasts. At first the car won despite itself, then it lost to the Ferraris and the Kurtis.

Sid Allard made some changes. First came the J-2X, the letter X standing for extended. The first J-2 was short in the cockpit and cramped the driver's

The Allard JR was the last and best of the sports-racing Allards, with much improved suspension, less weight and less bulk. Problem was, the other cars had improved even more. Road & Track

The Cunningham C5R was a smoother and sleeker roadster, still big and still with Chrysler V-8 power. It placed third at Le Mans in 1953, the team's best showing. Road & Track

legs. So the X got 6 in. added to the front of the frame (not the wheelbase), and the engine was moved forward 7.5 in., along with the transmission and the controls. The extra legroom translated into more weight on the front.

The J-2 was tail heavy and tail happy. The J-2X handled better but it didn't handle well, and the lumpy body didn't help.

Allard came up with the JR. This was smaller, with a 96 in. wheelbase, and had a four-tube ladder frame with torsion bars, de Dion rear suspension and split-axle front suspension. The body was sort of slab-sided but cleaner than the J-2 series and had less frontal area, which meant more speed. The standard engine was Cadillac, 329 ci, with a 9:1 compression ratio, an extractor exhaust and four twin-choke Solex carburetors. The engine was rated at 270 bhp, claimed dry weight was 2,200 pounds and the JR was as fast as the figures would indicate. The team entry led Le Mans for the first lap in 1953, only to retire later with a broken brakeline.

The Allard factory turned out 200 J-2s, 100 J-2Xs and either seven or ten JRs, depending on whose history you read. The JR was better, but it wasn't as competitive as the brutal old J-2 had been. Allard made some road cars and prudently retired into presenting England with the American sport of drag racing. That's still going on, while Allard the nameplate is history.

Cunningham C5R

Refinement comes in many forms. One might wonder if the airport venue influenced other forms of racing: Kurtis used live axles at Indy and for his sports car, as that form of suspension worked on flat, smooth surfaces. And in 1953 Cunningham went to the same concept.

The details were different, of course. The Cunningham's axles were located in a different manner, and the C5R was very close in other details to the earlier C4: Chrysler V-8 rated at 310 bhp, fully enclosed and street-practical body, and so on. The C5R's wheelbase was 100 in., tread was 55 in. front and rear, and dry weight was listed as 2,950 pounds. (No Cunningham ever earned the adjective *light*.)

There were signs of rethink. The transmission, for instance, was well behind the engine for improved weight distribution, and the brakes were monstrous 17 in. diameter Alfin drums—this with 16 in. wheels.

The C5R was clocked at 155 mph at Le Mans. Speed was fine. Thing was, although the Cunninghams had vastly improved teamwork and preparation, the Jaguars had that *and* disc brakes, and the Jaguars won. Cunninghams were third, seventh and tenth in 1953, and the same basic C5Rs were third and fifth in 1954, as well as taking their fourth successive class win.

It's worth noting here that the Cunninghams were real road cars, in that they were enclosed and had amenities such as lights and luggage space, even when the rules didn't require it.

Speaking of rules, there were three sets, not even worrying here about the various smaller clubs. The FIA set the course for the rules, which in 1953 were that a sports car, special or production, had to have lights, horn, starter, spare wheel and tire, with two

The hot rod wasn't dead yet. This is Duffy Livingston in his 1927 T roadster, using flathead Ford power. The car is just as it would be driven to, if not on, the dry lakes except that Livingston has a spare tire and is racing at the Palm Springs airport in 1953. Dean Batchelor

seats side by side and available for occupancy. There had to be one door, and dimensions were given for the size of the door as well as for seats and windshields.

The second set of rules, more of a dream than a framework for most builders, was laid down by the organizers at Le Mans, where such timesavers as cycle fenders were banned. The race had become famous as a test for production cars, so now the club took care to keep the pretext going.

The third set of rules, the ones that mattered most in the United States, were from the SCCA. In bare outline they conformed to the FIA's idea except that the SCCA was a member club, a nonprofit club run for the benefit of the people who joined, so the SCCA made an honest try at keeping the costs down and the older cars eligible.

The SCCA was being pushed into the national arena as the AAA (the competition branch of the national automobile club) didn't like road racing. The earlier experiment with shared responsibility hadn't worked. *Road & Track* editorialized that the AAA defended oval track racing because it made money on it, and because road racing "shows up the notoriously poor roadability of American stock cars." The AAA, said *Road & Track*, was mostly an insurance company and thus was forced to side with and defend the domestic industry. That may well have been true. However, in 1953 100,000 imported cars of all vintages were registered in the United States. It's hard to imagine how the domestic industry was threatened.

On the other hand the big Ferraris and the new Jaguar XKC ("Best sports car in the world," said *Road & Track*) were winning races. It was a Jaguar at Kimberly AFB, a Ferrari at Chanute in Kansas, another Ferrari at Lockburne, Ohio, with most of the competition coming from Stroppe's incredible Kurtis-Merc, which had four wins in its first five starts. The Cunninghams were home again, and doing their usual good work, except that the big event at Watkins Glen went to a newcomer named Walt Hansgen, in his own Jag-powered special.

Hansgen Jaguar

This was something of a new twist. Hansgen, an enthusiast who worked his way into the sport as a fan

Even more rare was this obviously homemade special, at Watkins Glen in 1954, powered by a Chevrolet six. Driver-builder Addison Austin didn't finish, but the C-Type Jaguar next to him won. Watkins Glen Racing Museum Archives

and later became a dealer for business and a driver for fun, simply took his own XK120, a good production racer, and stripped the body. He replaced it with a lighter, if less streamlined, aluminum body, sort of like a C-Type Jaguar in back and a Cunningham in front.

This was a good move. Jaguar was then in the forefront of production sports car racing so there was a good supply of equipment for the double-overhead-cam Jag engine, and the factory's own racing program, with the production-based C-Type, meant there was brake and transmission improvement as well. What Hansgen (and others) did was simply use the factory's development program to build backyard replicas.

Hansgen's car was fast, if not as fast as the specials with domestic power. It was also light, usually lighter than the V-8 specials.

Hansgen was an aggressive driver. Briggs Cunningham said later that while one can't compare drivers from different eras in terms of sheer skill and natural talent, it is fair to assign them certain characteristics: Phil Walters, the old pro, was a driver who always knew the car's limit and seldom exceeded it, while Hansgen went beyond the limit in order to learn it.

Hansgen's view, however, was another lesson. He had taken a new step, in using a sports car as the basis for a special instead of using a sedan, or building from scratch. But while his homemade C-Type was, with his

This begoggled warrior is Max Balchowsky and the hot rod is his not-terribly-tidy 1932 Ford roadster, with Buick V-8. Watch this space. Road & Track

Bill Rutan ran the Mount Washington hillclimb with his chopped and stripped Volkswagen Beetle. Road & Track

The best Chevrolet engines came from GMC in the form of truck engines, so if you just had to race a six, GMC was the way to go. This clean example of homemade sport was owned and driven by Chuck Tatum and used a GMC engine with aluminum body by Jack Hagemann. Dean Batchelor

help at the wheel, competitive with the factory's racer, in the long run he reckoned the money spent on his special during its time as a contender would have paid for two factory-built C-Types.

The real production racers were really good. Ferrari had discovered a lucrative market and the big red cars were winning, east and west. At the same time, early 1954, some attempts were made to widen the sports car market. Chevrolet had introduced the Corvette and Ford was hard at work on the Thunderbird, both more personal cars than sports roadsters in their first versions.

Sterling Edwards, last seen with his winning little Ford V-8 special, was now preparing a road sports car. Master mechanic Phil Remington recalled there were four versions of the road car, with either Ford or Henry J frame and with Lincoln or Cadillac V-8. The body, fully enclosed and with all comforts, was of fiberglass. Remington says that Edwards and crew got help with that material from the people at Glaspar, who in the early fifties were building boats but who later went into the kit car field.

Road racing—OK, sports car racing—had gone national. March AFB in southern California was the scene of the first true national race, in that the first five drivers in the main event were from back East—assuming Kansas City, Missouri, home of class C winner Masten Gregory, is the East. John Fitch and Briggs Cunningham were first and second in the Cunninghams, followed by Bill Spear. The western guys had mechanical problems—for instance, Troy Ruttman's Kurtis threw a rod—and some of the better West Coasters weren't entered. It was also a sign to see Sterling Edwards in a Ferrari.

But that didn't stop Miles from winning the smaller class.

Nor did the financial success of the SCCA-SAC link deter criticism. Some said that the runways were too flat and featureless, that the fans were kept too far from the track to see the action and finally that racing on runways wasn't the best way to improve production cars. Surely that last was correct, and one must be moved by the idea that racing was still expected to improve the breed.

There were other issues. First, Bill Spear, the SCCA's high points man in 1953, frankly said the award wasn't fair. He'd collected the most points, he said, in part because he won races but also because he was one of the few amateurs who could afford to race coast to coast.

Second, *Road & Track*'s letters column was enlivened by the old debate over sports car versus hot rods. The editor recommended "understanding between the two factions" and promised that the magazine's columns would be open for more debate. No fool he, as the readers loved the fight.

At about the same time, western racing was entertained by Max Balchowsky, who'd taken his 1932 Ford hot rod and filled the engine compartment with Buick. It wasn't the fastest car out there—it even lost to Miles when the Number 50 MG won overall—but it proved that the spirit of hot rodding still lived.

And third, the term dual purpose, describing cars that were built for the road but could be raced, in contrast to cars built to race but meeting the letter of the rules, appeared in print. This one was answered in part by a call for more careful inspection of specials, to be sure they conformed both to the FIA's rule book and to the laws of the state of registration.

In mid 1954, there was serious trouble. The deal between the SCCA and the SAC was under official criticism in the form of congressmen who questioned the public benefit from the use of what amounted to public lands, and the use of military personnel at private, bordering on profitable, events. The first bulletins said the series had been canceled. "Where do we race now," *Road & Track* wondered. "Are we racing for ourselves, or for the crowd?"

Eyerly Crosley

Some of us raced for ourselves. While most of the small engines came from overseas—witness the Miles and Barlow specials—there was another breed of homebuilt, running in the smallest SCCA class and equipped with an odd engine.

Back in the Great Depression, industrial magnate Powel Crosley, Jr., decided that what the public needed was a small car. A really small car. So he went into business with tiny sedans and station wagons, powered by a unique little inline four-cylinder engine, with overhead cam and (at first) a welded-up engine block and water jacket. The engine was almost too good for an economy car, and saw service in boat racing and in the military.

The Crosley car never sold well and the company melted away. (As did the idea of economy cars: One reason the domestic makers were so hesitant to downsize during the gas crunches was that they'd seen several times already that the public wouldn't put money where the critics put their mouths.)

The Crosley had a legacy, though, in the form of the engine, which was kept in production for years after the car disappeared and which was the engine of choice for home craftsmen interested in the smallest class.

Harry Eyerly was one such. He was a hydroplane racer from Oregon when sports car racing arrived. Cars looked like as much fun as boats, so he used his Crosley experience to build a mildly modified engine: redesigned carburetors, compression ratio, aluminum flywheel and steel billet crankshaft but stock camshaft because none of the racing models delivered better power in the right place.

The chassis was mild steel tubing, laid out in three rectangles. The front rectangle carried the engine, the middle section was for the driver and passenger seat, and the rear was for the rear suspension. The platform was light and stiff, with live axles

Harry Eyerly's Crosley Special used all the stock parts practical, and kept the metal body simple by avoiding compound curves. Road & Track

Eyerly's little car was so nicely proportioned that you don't realize just how small it was until you see it next to a loomingly large MG. Road & Track

and leaf springs at both ends. All practical parts, as in steering and gearbox, were Crosley. The frame was covered with a basic body, sheet alloy bent and hammered over a light framework and with no compound curves. The result looked something like a racing Jeep—or a shoebox on wheels, as one critic scoffed—but it worked and it won. Eyerly campaigned the car from 1953 through 1957.

When it first raced, the Eyerly Special weighed about 1,000 pounds and had a top speed of perhaps 95 mph. Eyerly refined the engine, with a special Iskendarian racing camshaft and tuned exhaust, and drilled holes in every piece of metal with enough square area to accept a drill bit. By the time he was through the car weighed 750 pounds, or one pound for every cubic centimeter of engine displacement.

The car ran without mechanical failure through the 1953 season, virtually unchallenged in class. The same went for 1954, except that Eyerly made some basic additions and also used the car in public road events such as rallies, just to prove that it could be done. The car won because it was simple and because he cared for the machine.

PBX Crosley Special

Candy Poole's PBX (*P* for Poole, *B* for partner Bob Bentzinger, *X* for experimental) Crosley was in the East what Eyerly's car was in the West.

Poole began with a Crosley Hotshot, the clever little roadster that the car company introduced just at the beginning of the sports car era. It was like an MG in that it had a lighter and smaller body with the same engine and running gear as the parent sedan, and with such features as removable lights and windshield and doors, for racing. Ahead of its time, as they often say when something doesn't appeal the way it was expected to.

The Hotshot was cute, but the chassis wasn't up to the engine and the engine needed understanding. The car had potential, in the overhead camshaft and the construction of the cylinder block and head in one unit, but the early versions of welded sheet steel were troublesome. Poole (and Eyerly) used the later cast-iron engine, which was stronger and more reliable.

Poole's car had a unique starting point, a 1949 Fiat 500 station wagon. The chassis was dismembered and the wheelbase set between the Hotshot's and the Fiat's. The frame rails were boxed and the rear section Z'd (as the hot rodders used to say) by cutting wedges

PBX Crosley	
Wheelbase	82.5 in.
Length	na
Tread	46 in. front, 45 in. rear
Weight	1,000 lb.
Engine	Crosley CIBA
Displacement	750 cc, 45 ci
Claimed power	55 bhp
Top speed	118 mph est.
0–60 mph	na
Quarter-mile	na
60–0 mph	na

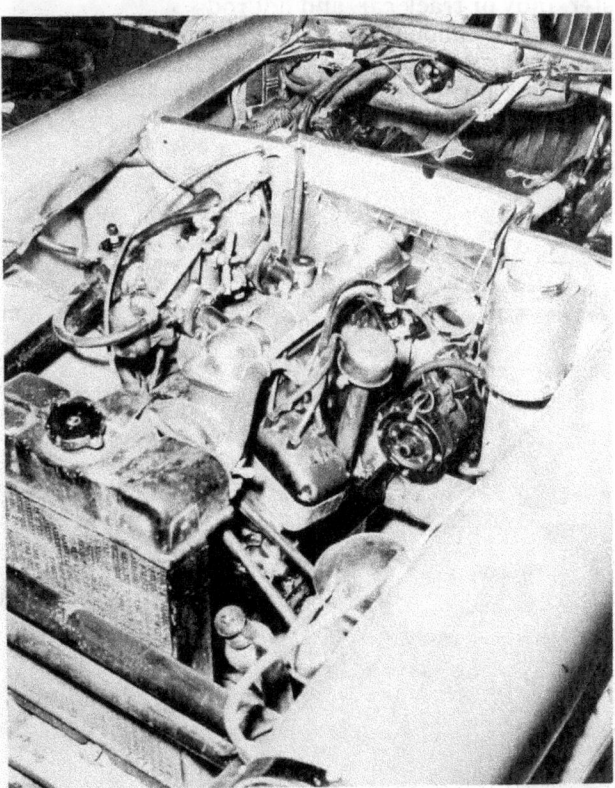

The single-overhead-cam Crosley engine in Eyerly's special used twin SU carburetors and was dwarfed by the magneto mounted on the cylinder head's left. Road & Track

The Crosley-PBX was the same size as Eyerly's car but had a much neater, professionally built body and used parts from Fiat as well as Crosley. This car was nearly unbeatable in its class on the East Coast. Road & Track

out and thus kinking the rails so the center was lower than the ends. The frame was as stiff as it could be, so Poole could use the Fiat suspension in front—transverse spring and wishbones—and keep it supple. This was radical in those days of flexible frames and stiff springs, but the approach was verified much later by designers like Lotus' Colin Chapman. Rear suspension was Crosley torque tube driveshaft and coil springs, with locating arms in the longitudinal and lateral planes. This allowed little wheel travel, so the rear was as, um, traditional as the front was modern.

You can talk with two tuners, both winning and both articulate, and come up with very different methods by which they achieve the same results. In this example, Poole spent most of his development time on the Crosley's camshaft, finally coming up with a set of timings and sequences that would let the engine rev beyond the normal limit. He also pushed the compression ratio to 10:1, and in fact once tried 12:1 but there was less power because the smaller compression chamber had the wrong shape.

Poole designed the body and made a scale model of balsa wood, which was shipped to a professional shop along with the chassis. The result was a much cleaner and neater shape. One rule was bent, sort of, in that the PBX's only door was a panel above the passenger's seat that hinged out; it was easier to clamber over the door than to use it—but the FIA never said the required door had to be practical, eh?

Sports Car Illustrated (now *Car and Driver*) did an analysis of the PBX. The car made thirty-five starts in its career, finished thirty-two times and won its class twenty-six times, along with beating cars in higher classes and even taking the 1500 cc race on a few occasions.

Eyerly and Poole were close in results, and it's a shame they never were compared in the heat of battle, or even in print. There were some quirks, in that Eyerly quoted his engine's power as 27 bhp delivered to the rear wheels, against Poole's 55 bhp at the flywheel with no accessories. Obviously the western car wasn't half as powerful as the eastern. If we compare weights, though, with the PBX one-third again as heavy as the Eyerly Crosley, we can guess that the Eyerly engine wasn't as powerful or as stressed, which explains the Eyerly's better finishing record.

Or, as has happened in racing, both teams were fudging on their power and weight, but in different directions.

Excalibur

Sports cars came from all directions.

Brooks Stevens was an industrial designer who worked with cars as well as consumer products. He noticed early that people were interested in cars as fun, as well as for racing, and announced his own sports car, the Excalibur.

It wasn't new ground exactly, in that the base equipment was normal sedan, but Stevens used the Henry J from Kaiser-Frazer for the chassis and picked an inline six, the 100 bhp Willys, for the powerplant. The body was distinctive: a blend of early open roadster, Indy or track car, and hot rod.

The Excalibur didn't work. There was a factory team, two cars strong, for racing, but the car had 100 bhp, had a wheelbase of 100 in. and weighed better

The Excalibur, shown here as a racing team in the make's early days, was built and assembled from unusual sources, as in Kaiser-Frazer and Willys. It was a little too bland for racing, a little too Spartan for touring, so it never went far. Road & Track

than one ton. So while it was gallant, the Excalibur didn't win races.

There was talk of buiding production versions with more exotic engines, but the lack of interest in the basic project convinced the builder-designer that he should do other things. As it happened, the name was kept on the shelf after the sports car was abandoned, and Stevens' firm did quite well later, using the label on rough copies of the classic Mercedes-Benz touring car.

Sports car racing was wildly popular. The threat to the SAC contract eased up some, and there were still public parks and so on, and a few tracks, so the sport flourished. One result was strong interest from Ferrari, who had perfected the big semi-production racing roadster.

At San Francisco's Golden Gate Park, 100,000 people watched Chuck Daigh lead in a Kurtis only to break, then Stroppe did the same and Jack McAfee took the prize for Ferrari. Elsewhere, Jim Kimberly took nine of his ten races that year, also for Ferrari.

Tracy Bird was then building and racing his own car but would go on to become a power in the SCCA and in racing. He wrote to *Road & Track* that money was the main reason people built their own cars, because "most of us who have specials would give our eyeteeth for a real Ferrari."

By this time, the imported pure racing car, as in Ferrari and to some extent Jaguar, had a big engine. Formerly there was a choice of small racing engine or large sedan (domestic) engine, so the extra weight of the big job could be compensated for with the extra power. Now it wasn't so easy.

Cunningham's team won Watkins Glen with the tried-and-true C4R, but more often than not the more tried-and-true flathead Ford was faster although it blew up under the strain.

Troutman-Barnes Ford

The pure skill and determination of the guys who were still building and driving the flathead V-8s kept the ancient design in contention. For example, Dick Troutman and Tom Barnes were skilled metalsmiths, in aviation and later in Frank Kurtis' shop. They wanted to get into road racing, they knew they couldn't afford a Ferrari or Maserati or whatever, and they also knew they could build as well as anybody in the world.

This was not a homemade car. Troutman and Barnes used a short wheelbase of 87 in. so the car would react quickly. The frame was steel tubing and the body panels were aluminum. There was lots of that amateur's dread, the compound curve. The car was clean and slick, if not especially pretty—the sort of car that the phrase "gets the job done" was created to describe. The Troutman-Barnes car had alloy wheels and knock-off quick-change hubs from Indy and sprint car practice, MG steering, and Ford engine and transmission. It was a mix not quite of old and new, but of what would do and what had to be first-rate despite the cost. Clever application of transverse springs gave the car enough suspension to work on the smooth airport and park surfaces.

Another advantage the home team had was simply the ability to build for the course. The big Ferraris and Jaguars were always intended for racing,

Chuck Daigh is at the wheel of this Troutman-Barnes Ford Special at Golden Gate Park, California, in 1954. The Ford V-8 was outmoded but not outclassed, and Troutman and Barnes had a good combination of high tech and simplicity. Road & Track

Troutman and Barnes were craftsmen and their hand-formed aluminum body shell was nearly perfect. Taillights were Lincoln. The license plate means the car could have been driven on the road. Road & Track

Cunningham team driver Sherwood Johnson had a hobby, so to speak, in the form of this Jaguar-powered special. It was basic, even crude in places, but Johnson held the Mount Washington hillclimb record in 1954. Road & Track

but they were required to run the classic events—Le Mans in France, Targa Florio in Sicily and the like—as well as thirty-minute dashes at the neighborhood air force base. So they were overbuilt, on purpose. When Ak Miller came home after his Mexican adventure it took him only one race to learn that his endurance racer was too heavy for SCCA's sprint racing. He sold the car and built a lighter and simpler one for United States events.

Troutman and Barnes had learned all this from Kurtis before they began, so their car was light, just heavy enough to last the weekend, and they could stress their flathead V-8 to the maximum because they knew precisely what the maximum was.

If that wasn't enough, and it wouldn't have been, the team had driver Chuck Daigh, who was both a natural talent and a skilled technician in his own right. He could drive, and he could translate what the car was doing and thus was able to tell the crew what had to be done next.

Devin

Just when things looked their worst in the engine department, help arrived from another quarter.

Bill Devin was a racing enthusiast, and a man skilled in the handling of fiberglass, which in the early fifties was being hailed as the material of the future. Chevrolet was ready with the Corvette, the boat crowd had begun to rely on what the wood-boat defenders referred to as "frozen snot" and the mechanical magazines were fulsome in their predictions of the mass adoption of fiberglass for cars.

In general, it didn't happen.

But in one corner, Bill Devin had a great idea. He borrowed a lovely Italian racer, an Ermini roadster, and made an exact copy of the lithe and curvaceous and efficient body. He took a mold of the car from the outside—just laid the material atop the car, and when it cured and was lifted off, *presto!*, there was the body's mirror image. All you had to do was apply fiberglass cloth and resin and catalyst in the correct proportions and you could lift out a shell, ready to be cured and finished and painted and mounted on the frame of your choice.

Devin got into the business as a sport, so his project did him some good in the short run; he wasn't a top driver or tuner. His small step for one man, though, became a big leap for everybody else who was building, or thinking about building, a special at home. A quick review of the homemades shown thus far will establish that the builder either went to a profes-

No, this isn't a special. It's an Ermini, an Italian sports racer, and it's the very car Bill Devin borrowed so he could make the mold from which he produced his fiberglass bodies. Road & Track

Devin's lovely copy came in handy. This example of the art form is a Panhard, a front-drive French car, with the body shell suitably modified. Road & Track

sional fabricator or stayed at home avoiding compound curves.

Devin's body could be adapted and adopted, revised and reworked to fit virtually any frame or chassis or floorpan; it went with front- and rear-engine cars. Devin went into production of the body shell and of the various bits (such as instrument panels and seats and hardware) that a builder would

Pete Lovely and his Porsche-Cooper, probably at Pebble Beach, and almost surely having trouble with the throttle linkage. Road & Track

The Pooper's Porsche engine did some serious modification to the Cooper's bodywork—not that the car was an esthetic treat to begin with. But it met the rules of the time and place, and the combination worked perfectly. Road & Track

need, and could buy cheaper and easier than he could revise from whatever stock car he had begun with.

This was a big advantage. When Ak Miller built a lightweight car for local events and needed a body, he did the sensible thing and got a Devin. Miller was one of several hundred racers who took advantage of the simplified body; it's hard to know exactly how many did so because the completion rate was low. For every person who got a Devin-bodied machine on road or track, several abandoned the project when they had to admit that it was just too complicated for them.

Devin wasn't the first man who saw the advantage of fiberglass, nor was he the first to produce and sell bodies of the material. Instead, he did the job best.

The Pooper

Pete Lovely was a Seattle, Washington, Volkswagen-Porsche dealer who liked racing and liked doing things his own way. He was also a good, fierce driver.

Cooper was an English brand of racing car. The company began with tiny single-seaters powered by 500 cc Norton Manx motorcycle engines, back when the purists believed in keeping costs down. (The purists have always believed in that, and a dozen sets of rules have been drawn up to keep costs down. They haven't worked, ever.) Cooper would eventually revise the world by proving that in formula cars rear engines are better than front engines and that small beats big, but in 1954 the firm was expanding just a bit and offered a sports car.

Lovely's Pooper	
Wheelbase	87 in.
Length	na
Tread	44 in. front, 44 in. rear
Weight	920 lb. dry
Engine	Porsche 1500 S
Displacement	1488 cc
Claimed power	84 bhp
Top speed	125 mph est.
0–60 mph	7.0 sec.
Half-mile	15.0 sec. e.t.
60–0 mph	na

Sort of. It was really the single-seat car except that it had a body that covered the wheels and provided a place for headlights, and it had a second, smaller seat. The main seat, the driver's seat, was in the center of the chassis and the passenger went on the driver's left. Remember that.

At the time, Porsche offered its 1500 cc flat four in road, sporting and racing versions. Lovely took one of the milder ones, the 1500 Super, and put it into the little car bone stock, on the theory that he wouldn't need more power than the engine already had and that he could ensure reliability if he didn't change the engine.

Road & Track tested the Pooper—the magazine couldn't resist the mixed and matched names—in 1956, after the car had been winning for a season and was national champ in class. The testers were impressed. The Pooper was a match for the OSCA 1500, which was the standard in the under 1500 cc class then, from a dead stop up to speeds beyond what the little cars would hit on airport or park circuits.

There was no question about what the Pooper was supposed to do. It was a fine car for racing, period. The passenger was an observer. *Road & Track*, which always tested with driver and reporter on board, commented that getting into the second seat "takes a contortionist." The magazine went on to say that "the seating and door meet all current technical requirements but are obviously an expedient and not designed for practical use." Just so.

Otherwise, the Pooper was easy to drive, once one got used to a seat that leaned *way* back. There were no vices and cornering power was far beyond what the testers, trained as they were in road machines, expected.

Road & Track again: "Lovely's beloved little Pooper is certainly one of the most interesting specials now extant, and its success should do much to

Beneath the skin, the original seating position for the pilot and a second seat, to the driver's left, for the rule book. The fuel tank is in front, for balance. Road & Track

encourage other special builders who either cannot afford an expensive competition machine or merely want to have the fun of designing and building a car of their own."

Miles Shingle

When last seen, Englishman Ken Miles was routing the establishment with his torpedo-bodied MG. The car won and then was eclipsed, so during 1953 Miles took notes and planned his next project.

Old Number 50's main strengths had been good handling and better reliability. Its weakness was simply lack of speed, especially in view of the later Porsche and OSCA racers.

The MG engine was Miles' only choice, he wrote, and it was stressed just enough in the state of tune already achieved, with 10.7:1 compression ratio and revved to 6800 rpm. The next-best hope would have been to keep the engine the same and make the car lighter, but lightness costs money and Miles' budget was probably limited since MG was established on the market by then and so he got less help from his employer.

The remaining avenue was to go faster with the same weight and power. To that end, Miles laid out a multi-tube frame, an early effort toward what would later be called a space frame, of small-diameter steel tubing. The smaller tubes let the driver be next to, rather than atop, the frame rails. Miles went on to use such tricks as making the oil pan higher and wider and tipping the engine so the crankshaft pulley was the same height as the bottom of the flywheel. Tiny steel stringers were wrapped around the contours of the wheels, the engine and the driver, and sheet aluminum was wrapped around them.

The body's shape, by no coincidence at all, resembled the then-current MG TF's, with separate fenders and grille, except that the Shingle, as the car came to be called, was perhaps half as high. This was radical stuff, but with a cut-down MG grille and paint in traditional British Racing Green, nobody had any trouble telling whose side Miles was on.

The Shingle demonstrated other new thoughts, as in the tuned exhaust and the revised intake manifolds that put the carburetors lower than the exhaust pipes. The car was a fine piece of work. It weighed less than Number 50, by a few pounds, and surely had an extra few horses if only because the builders and tuners had had another two years to work on the thing. The new special shared the stock-based ideal of the old, except that this time the Morris and MG steering and suspension were further developed.

Alas, things had progressed elsewhere as well, and the new car wasn't as much better than its rivals as the earlier, more basic Miles special had been. And there was another sort of retribution: Miles won his first race in the new car, but then was disqualified on technical grounds.

Always popular with the troops, Miles wasn't fond of the officials in the California Sports Car Club, a feeling returned in full measure. He was elected president of Cal Club, then re-elected for a second term in 1955, only to have the people who ruled on such things say that the Shingle wasn't legal because the rear tires we

That was easily taken care of. The next problem was that the old pushrod engine, with its long stroke and spindly connecting rods and flimsy crankshaft, couldn't keep up.

Yes, there is an answer to the obvious question: At the national SCCA races in Sacramento, California, the Lovely Pooper beat the Miles Shingle. By six seconds. In some classes the old reliable could beat the new upstart, but not in others, including the 1500 class.

Miles was so obviously a top driver that he went on to get rides for the MG factory and with Porsche and with a series of specials (Troutman and Barnes, for instance) and then with Ford when that make ruled the racing world. He was killed testing for Ford in 1966.

Cunningham C6R

Another sad story. The original Cunningham plan was to build racing and road versions of the brand, and win races such as Le Mans, while getting some of the money back from the production cars.

It didn't happen. The racing effort was successful in the United States and overseas; the team did lead the race and the cars were as fast and as good as the competition, it was just that things broke at the wrong time and place, or the other guy had one more arrow in his quiver.

This went on for years, until it was clear that the Chrysler V-8 was too big and heavy for its potential power. Cunningham toyed with Ferrari but decided against using an imported engine for political reasons.

The next replacement was American to the core: the Offenhauser, a true racing engine with double overhead camshafts and so on, a staple of American professional (oval track) racing for a generation.

The Cunningham team decided to build a new car, using what they'd learned previously. Because the engine was smaller and lighter than the Chrysler V-8, the car could be smaller and lighter, too. The C6R (Cunningham followed a logical numbering system, most of the time) had a tube frame, independent front suspension by means of A-arms and de Dion rear axle, all clothed in the slickest roadster body in the make's history.

The Offy engine was a lot of work. It was a big, rough inline four, designed in an earlier and different era by bright men. It was a wonderful air pump. But it was designed when oval track racing cars burned alcohol and it was a racing engine so it didn't need all the extras found on passenger—and sports—cars.

The Cunningham team's first job was to fit the Offy engine with things like a generator and a flywheel

Ken Miles, looking for all the world like a man who's read the script of his own movie, shows how tightly his new special was constructed around him. The tiny wires are stringers for the body panels. If you look closely, you'll see that the engine is tipped down in front—just a fraction, but Miles never missed a trick. Road & Track

Artisanship, as in the engine-turned panel for the dashboard, was evident everyplace on the Shingle. Road & Track

The Shingle's exhaust pipes were tuned, with same-throw cylinders one and four paired, and cylinders two and three paired, and those two pipes joined, so the pulses would extract exhaust at chosen engine speeds. In addition, the hood line could be lower because the carburetors were mounted lower than stock, below the exhaust pipes. Road & Track

with a ring gear for the starter motor that also had to be installed. The Offy was tuned for alcohol, and that had to be changed to suit the gasoline used at Le Mans.

The adaptations proved the car's undoing. At Sebring the flywheel came loose. It did that because a sports car or road racing car needs to use a clutch all the time and a track car doesn't. The flywheel's mounting bolts got loose in their holes under the stress of torque being applied and released.

Le Mans was a disaster for the world. There was a collision between two of the racers and flaming parts shrieked into the crowd. One driver and seventy-six spectators were killed. It was the worst racing accident since the Paris-Madrid debacle of 1903. Almost unnoticed were the air brakes of the Mercedes, the victory of the Jaguar or the retirement, due to a broken connecting rod, of the lone Cunningham.

The Offy engine was a logical and perhaps sentimental choice for special builders, but it never worked as well as its track record promised. The reason seems to have been the choice, or lack of choice, of fuel.

Miles skillfully reworked a grille from the then-current MG TF so it fitted perfectly and blended with the front body panels, so there'd be no doubt as to what brand of car this was supposed to endorse. Road & Track

Beautifully built and developed, the Shingle was a work of art, even if it wasn't as much better than its rivals as Miles' earlier special had been. Road & Track

The Cunningham C6R was smaller and lighter than the previous models, but like them was built mostly to race at Le Mans, in America's colors. Road & Track

When Cunningham's crew was prepping the Offy, they ran extensive dynamometer tests. The engine would instantly crank out 270 bhp, which was good from 180 ci or three liters, but after a few minutes power would slip away. The water jacketing around the valve seats had been designed for alcohol, which runs cooler than gas. Evidently the evaporative powers of alcohol had been keeping the valve seats cool, and solid, all those years. But gasoline didn't cool the seats and they became soft and flexible, and the valves didn't seal well and down went the power curve. Easy to describe, not so easy to cure.

Cunningham had gone as far as he could go along this trail. In the beginning there were race and road versions, with the former built at Cunningham's shop and the latter bodied by Vignale in Italy, but only one model, the C-3, was truly produced and sold to the public. The demand wasn't there and the money wasn't there.

Late in 1955 the Le Mans organizers announced some new rules. The production class would require 100 examples, while prototypes were limited to 2.5 liters of displacement, supercharged or not. The intent surely was to slow down the cars and demonstrate to the French government and the public that the club was concerned with the safety of the spectators. The incidental result, though, was to force Cunningham out of his role as producer of American sports cars. He and the team would race for years to come, but they'd do it with teams of Jaguars and Maseratis.

(A footnote: When the Cunningham team modified a Ferrari, the designation C6R seems to have been

The rear suspension of the C6R was de Dion, with the wheels independently mounted and the driveshafts (halfshafts) articulated, but with the rear hubs mounted on a tube that ran from side to side, behind the differential. The differential and the rear brakes were sprung weight, which helped handling, while the wheels were in perfect alignment. It worked, but it was complicated and expensive. Road & Track

The entire C6R was complicated. There were rear and side fuel tanks, with the oil tank above the firewall on the driver's left and with a cooler to the left of the radiator. The twin-cam Offenhauser engine had been fitted with starter and generator, but couldn't be successfully adapted to run on gasoline. Road & Track

The Kurtis 500X was an improved version of the more successful original. It used live axles and torsion bars, but the frame was multi-tube, lighter and stiffer. The streamlined body gave more speed from the same power, while the power—Buick V-8 in the best example, driven by Bill Murphy—was increased. Even so, fewer were sold and they won less often. Road & Track

used, then discarded. According to later factory records there was only one C6R, the 1955 team car.)

Kurtis 500X

The handwriting was on several walls. The Kurtis 500S had been eclipsed by 1955, perhaps because Stroppe was forced to sell his Kurtis in the wake of partner Clay Smith's death, and the other Kurtis drivers weren't the demons he was. Ferraris and Jaguars were doing the winning.

Frank Kurtis introduced a replacement sports racer, the 500X. It was in the same mode as the 500K, being a wider version of the Indy roadster, with live axles front and rear. However, the 500X's frame was a semi-space design, Kurtis being as able as Ken Miles when it came to seeing what was an improvement. And the 500X was streamlined, with a fully enclosed body, aerodynamic lines and carefully sited openings. It was several hundred pounds lighter than the older car and came with almost the same choice of engines—well, by that time no company had the Hudson six in its catalogs.

The 500X was clearly a better racing car than the 500K. But times had changed. Bill Murphy, a California Buick dealer, and Bob Schroeder, a Texan, were successful in their examples, but only twelve 500Xs were made. Kurtis made a handful of road machines, 500Ms, then faded back into secret prototype work for the government.

Millikin Four-wheel-drive Butterball

This is just for fun. The rules for this treatise say the cars must be sports cars, with two seats and so forth; this wasn't.

The Butterball was a combination of Jeep driveline, German Steyr V-8 tank engine and single-seat body. It was air-cooled and had eight carburetors and too much power. It didn't win, but it did make an impression. Road & Track

Bill Millikin was an early enthusiast, an engineer and one of those who made Watkins Glen a success. He also named a curve there when he tipped over in his Bugatti.

Soon after the end of World War II, Millikin was in England doing work for the military and he came across an odd special. It was built by an English racer, Archie Butterworth, and it was made from combining parts from a four-wheel-drive Jeep with a captured German Steyr V-8 tank engine. The Steyr displaced 270 ci and had a compression ratio of 14:1. It ran alcohol benzine through eight tiny carburetors, one per barrel, and it was ferociously powerful. This odd couple was covered with a single-seat body, kind of like an American sprint car.

Millikin fell in love. Butterworth wasn't interested. Millikin went home, only to get a letter from Butterworth. He was in the hospital. Seems he'd gone into a turn too fast and been pitched out and whacked by the roll bar of his own car. "At that point," Millikin recalled, "he was willing to sell."

The car arrived crumbled. When they took the thing out of the crate, someone quipped that it was made by Butterworth and came wrapped into a ball, so it should be called Butterball which it was ever after.

With hundreds of horses, swing axles, four-wheel-drive and the tires of the day, the Butterball was a handful. Early in SCCA history, it ran in the unrestricted class. Then came hillclimbs, including Pikes Peak in Colorado. The idea was ahead of its time, or perhaps ahead of the rest of technology, but nobody who ever saw it forgot the sight.

Morgensen Special

This is a preview of things to come, and come again.

In 1953 Dick Morgensen built a special in his barn in Phoenix. According to legend, he welded up a tube frame and then installed whatever he laid his hands on: a Plymouth six for an engine, the front end of a 1938 Ford, rear fenders from a Chevrolet truck and so

We're at Santa Barbara Airport, 1954, and we're looking at the special Dick Morgensen built from leftover parts, including the side-valve, six-cylinder Plymouth engine complete with three carburetors. Not, one would think, a candidate for national fame. Dean Batchelor

Now it's Willow Springs, 1955, and the Morgensen Special appears to have a V-8 with four carburetors, and to be leading Ken Miles' Shingle. When Road & Track *used this photo, the Morgensen car was cropped out of the frame.* Road & Track

on. That doesn't sound believable, until you look at the car in its early form.

Morgensen won a few races and placed high in a few others. He brought the car to California and left it, for sale on consignment, in the shop of Hollywood mechanic Max Balchowsky.

Long's Renault

Here is another preview, but this time in theory.

Al Long worked for Meyer-Drake, builders of the Offy engine, and he raced midgets right after World War II. When sports car racing arrived, he parted out a 1948 Renault sedan, which came with a rear engine.

Al Long, in his pioneering Renault Special. Long worked for Meyer-Drake and was a skilled fabricator—witness the shaped metal body. Road & Track

He combined parts of the Renault frame with steel tubes and reversed the engine, so it was within the wheelbase rather than protruding aft of the rear wheels. The 750 cc engine was mildly souped up. The body panels were aluminum and steel; weight distribution was thirty-five percent front, sixty-five percent rear; and estimated weight with all parts was 826 pounds.

In the event, this car didn't set records. But it did show that people were thinking about where the components would work best.

Morris Minor Special

This is an example of the great-oaks-come-from-little-acorns school of construction.

The builder was Forrest Edwards. Edwards liked superchargers. He had one on his MG TF and it worked fine, so he put one on his 1951 Morris convertible, which came with a 900 cc side-valve four, rated at 27 bhp. The little car was suddenly much faster, powerful enough to stay with stock Jaguars for the first block or so, which inspired Edwards to remove the mechanical parts and throw the Morris body away.

Edwards built his own frame of steel tubing, to which he attached the Morris rear axle and front suspension and steering. He used stock transmission and a beefed-up clutch.

The engine was treated to the normal improvements for a supercharged powerplant. Compression

Long used Renault cross-members and front suspension and chassis parts, with part of the frame made of steel tubing. The rear engine was swapped so it was within the wheelbase and the fuel tank was mounted at the right front of the chassis, to balance the weight. Road & Track

A Morris-based special at Pebble Beach, 1954. The car used a supercharged side-valve four, a homemade frame and one of the best amateur bodies we've seen so far. Road & Track

ratio was lowered to 5.3:1, then a Roots impeller-type blower was mounted next to the engine and driven by a belt. The blower's maximum pressure was 14 psi, or close to one atmosphere.

The body was hand-hammered alloy; Edwards was a gifted amateur. There was one door, on the left, but the car was never intended for road use and had the headlights, which were tucked behind the grille,

When you needed big engines to get big power, you also had to build a big car. The X-Ray Special used a massive frame of large tubing, with a 275 bhp Chrysler Hemi V–8, and with gearbox, wheels, brakes and so on in scale. Road & Track

The X-Ray Special's body was futuristic for 1955; rear lights and fenders have a touch of Corvette about them. The car never got beyond the prototype stage, mostly because it was just too big for racing and too Spartan for the price a builder would have to charge. Road & Track

too low for legal highway recognition. The car made a few races in 1954, and had a good record, if not a brilliant one.

X-Ray

Power alone wasn't the answer. One of the better built failures was the X-Ray Special, the pet project of a wealthy industrialist.

The car had all the makings of a winner, including a full-race Chrysler Hemi V-8, rated at 275 bhp. Most useful, perhaps, would have been the sheer, pure torque of the engine, which was much larger than any Ford could have hoped to be.

Bill Baldwin had been taking notes since he built his sprint car-track roadster-sports racer. This Baldwin Mark II used a flathead Ford V-8 but was much more compact and trim. The wide scoop on the hood clears the top of the car's second radiator. Baldwin reasoned that two small radiators, one aft of and above the other, could result in a smaller total frontal area. Road & Track

The X-Ray was a big car, with a 99 in. wheelbase and a 54 in. tread. The frame was made of 4.5 in. diameter steel tubing. Rear suspension was de Dion, with locating tube and with the differential mounted solidly to the frame, just as the book recommended. Front suspension was trailing arms, similar in principle to the system used by the Volkswagen Beetle, except that the front and rear wheels of the X-Ray were sprung by transverse leaf springs.

The X-Ray got a good write-up in *Road & Track*, which ended the review on a confident note that any day now the big Ferraris were going to get theirs.

As they did. But not from the X-Ray.

The Chrysler Hemi was a wonderful engine in principle. It moved the domestic industry into the present. It worked in passenger cars and in drag racing for another ten years. But it was too heavy and bulky to compete in sports car racing by 1955.

Baldwin Mark II

Ferraris or not, in 1955 there was still a place for the carefully constructed and driven domestic. The Baldwin Mark II, so called because there'd been an earlier and similar car from the same shop, was one of the last semi-successful racers to use the venerable flathead Ford.

The Baldwin was a simple car, with the now-classic torpedo shape and cycle fenders and tacked-on road equipment. But some thought went into the use of two small radiators, one behind and slightly higher than the other so frontal area could be kept at a minimum.

The driver was Bill Pollack, who a few years earlier had been a terror in a J-2 Cad-Allard. (It might have been that drivers who could win with an Allard had been so thoroughly trained in control of panic that no car could ever frighten them again.)

The Baldwin was a good horse for the course. On the day it won the feature race at Willow Springs, California, Ak Miller was in the field in his Mexican road race hot rod, the Caballo de Hierro. Pollack's flathead special beat the Olds hot rod, humbler of Ferraris, by six seconds per lap.

Beavis Offenhauser

Cunningham wasn't the only man who looked at the Offy engine with speed in his heart.

Offenhauser made a selection of big engines, double-overhead-cam fours with varying bores and strokes to allow for the various jugglings of the rules, and it made a smaller version, for midget racing. By happy chance, the midget Offy displaced 1500 cc, which in 1955 was the dividing line between the big and the small road racing classes.

George Beavis did the logical thing and laid out sort of a big midget Offy. It had a torpedo body, like the single-seater's but wider, and with cycle fenders. There was one door, on the driver's side, in the shape of a rectangular panel that obviously was never going to be used as a door. The lights were bolted on and the required spare tire went atop the rear of the body. The body was made by a top firm, Autocraft, that did paneling for many West Coast oval track cars and for special builders who could afford the tab.

The 1500 cc Offy engine could be bought with clutch and four-speed gearbox, so some of the work done by Cunningham's crew wasn't required. The engine was good for 100 bhp, tuned for the day, which made it a contender. Except that the adaptations never worked as well as hoped, and the track engine was never happy on gasoline, nor was it as flexible as, say, the Porsche 1500 racing engine. The Beavis Offy won an occasional race, but that was all.

This was a period of change. The SCCA-SAC program wound slowly to an end, although there was an occasional event where the contracts were firm. Promoters with vision were building closed-road circuits—especially in towns like Elkhart Lake, Wisconsin, and Watkins Glen, New York, and Riverside, California, where races had been held and the town fathers knew what sporting events could do for revenue. Some of that would continue in an earlier form, as various cities bartered use of parks and industrial complexes as occasional race circuits.

What the old airports and the new racetracks had in common was smooth surfacing. The classic races were going away, and when the tracks changed from real road to artificial road, the demands on chassis and brakes and suspension were different. Not less, surely; the extra power from larger and more reliable engines would take care of that. But the different course was best conquered by a different horse.

The Beavis Offenhauser looked and performed like an enlarged midget track car, with the spare tire mounted on the rear deck and with that little rectangular panel under the driver's elbow posing as the SCCA-required door. Good work, but the Offy midget engine was unhappy on a gasoline diet, as was the larger Offy in the last Cunningham. Road & Track

Chapter 4

Sizzling Stovebolts 1956-1958

Just as in the Good Book, the humble get mighty and the mighty get humbled

Racing's happiest accident began way back. Call it the early thirties, when Henry Ford introduced his flathead V-8. He did so because his four-cylinder engines were being eclipsed by Chevrolet's six.

Ford's jump from six to eight could have inspired Chevrolet, or General Motors, to fight fire with fire. Except that it didn't. Ford was willing to spend money and learn techniques, even when the urge to excel forced it to invent foundry skills the rest of the world couldn't match.

GM had plans for Chevrolet so the bow-tie brand got independent front suspension and hydraulic brakes and lovely colors and pleasant interiors. But Ford was the hot one, to use an advertising slogan Chevrolet used for itself decades later. Clyde Barrow was kind enough to always steal a Ford when he could, and vain enough to drop Henry Ford a note to that effect, while Ford in turn was canny enough to let the press know.

There was no performance market then. There were no hot cars, no factory hot rods. There was a scattered fringe of nutball car guys who drove and raced and rebuilt and modified Fords. Chevrolet was the doctor's car, the brand your dad drove until he could move up to a Pontiac or Buick.

GM was strong on marketing and production rather than innovation. Its cars had some cheap touches, little places where cost was the object, and because of that Chevrolets were known as Stovebolts, a reference to cheap hardware.

There's no proof anybody minded. Chevrolet was the bottom of the line. When GM decreed new overhead-valve V-8s designed to take advantage of better fuel and higher compression, Cadillac and Oldsmobile got the first tries, in 1949. Then came Buick, and finally Pontiac and Chevrolet for model year 1955, a year later than rivals Ford and Mercury.

The Chevrolet V-8 was far beyond anything anybody had expected. It was small: 265 ci, compared with 331 ci for the Chrysler and Cadillac, 324 ci for the Oldsmobile. It came in two versions—rated at 162 bhp with two-barrel carburetor and 180 bhp with four-barrel carburetor—and at first impression was a nice little sedan engine.

None of the men who designed the engine ever came right out and said so, but surely they gave long

The basic Chevrolet V-8 was smaller on the outside and inside than the early Cadillac, Oldsmobile and Chrysler engines. It was also lighter and revved higher and breathed better than the Ford V-8, which got overhead valves one year before Chevrolet got an eight. Without this building block, racing wouldn't have been the same. Chevrolet Division

hours of thought and work to the racing potential of their engine.

First, it came with a short stroke of only 3 in. The crankshaft was strong and the bearings generously sized, and it was safe to rev the thing 8000 rpm or even 9000 rpm, when the other, longer-stroke engines came apart at 6000 rpm.

Second, the shapes of the ports and combustion chambers were simply... right. Air and fuel flowed in and out. The valves were of good size and in the locations, at the angles, that made good, calm power.

Plus, the valve gear. In the books, overhead camshafts are better than cams in the block working pushrods and rocker arms because the less gear between cam lobe and valve, the better the valve follows the contours of the cam. The less flex and wobble, the faster the engine can turn, and power comes from revs.

The Chevrolet was overhead-valve, sure. The difference was that where the earlier V-8s had shafts carrying heavy rocker arms, the Chevrolet engine used stamped rockers that pivoted on little ball joints plugged into the heads. The lighter and stronger valve gear took advantage of the strong lower end by delivering the potential.

There was a little problem: Between cam lobe and rocker were valve lifters, to cushion the impact of the lobe on the pushrod. The lifters were hydraulic. They used oil pressure to maintain this cushion, and if you revved the Chevrolet too high the lifters would pump up (that is, lose the cushion) and the valves would float, not follow the dictates of the camshaft. This made a terrible noise, and it did damage and you lost the race.

All this happened virtually in secret. When the V-8 was introduced, its sponsors assumed they had a good sedan engine.

There was also the Corvette. Introduced in 1953, it was a roadster in that it had side curtains instead of roll-up windows. It was called a personal car, an advertising agency's way of telling you it wouldn't perform any better than your neighbor's sedan. Ford was in the same mode, with the two-place Thunderbird of 1955. These were nice looking little cars, and you didn't have to actually drive them to enjoy them.

The pieces had barely been put into place when the stock car crowd discovered that the smaller Chevy sedan with V-8 made a natural short track racer. The street racers were running parallel, so to speak, with the Power-Pack V-8 in the 210 business coupe, the lightest car in the line-up and an easy winner of the the stock class at the drags, not to mention impromptu contests leaving the drive-in.

This happened just about the time the public took sports cars to heart, and just as drag racing—legalized and socially acceptable hot rodding—came into the public eye.

Performance was transformed into a mass hobby. For Chevrolet, it was as if the good guy had stepped onto Main Street, looked down and discovered, just as the bad guys loomed, that his mom had strapped on his six-guns when she straightened his tie.

Those who take notes may recall that when the sports car movement first took life, there was an overhead-valve conversion for the Ford V-8. The designer and maker was Zora Arkus-Duntov. When that project faded, Duntov went on to other things, culminating in his appointment as chief engineer for the Corvette.

It was a long and fruitful relationship. Arkus-Duntov was a sportsman, a skilled amateur driver and racer, and he was among the first to know what happened to the valves when you revved the engine too high. The rest of us made faces; Arkus-Duntov made a new camshaft. The Duntov cam used solid lifters and got more power and you could rev it till the tach needle bent. Wonderful stuff.

Chevrolet's marketing department was quick to cash in on the serendipitous appeal of the new engine and the engineering department was happy to cooperate. The Corvette team became a real team, with real racing cars and with parts and advice for private owners. The SCCA was working into a fair system of production racing at this time and was happy to have the Corvette, Made in the USA and all that, at the top of the heap.

What's all that got to do with specials?

Ready for installation, the Chevrolet V-8 weighed 535 pounds. The Chrysler weighed 750 pounds, and the Caddy and Olds each tipped the scales at 710 pounds.

Imagine how many lightening holes you'd have to drill in frame, brake drums and so forth to get shed of 200 pounds. Even then, the heavier engine needs a stronger frame, needs bigger brakes, uses more fuel and so on. Everything has to be in scale, so a car with 300 bhp from 300 ci can be lighter and thus faster than one with 300 bhp from 400 ci.

In this light, it's perhaps surprising that every backyard builder in the United States didn't switch to Chevy power on the first of the year.

But they didn't. The new V-8 was shown to the press in the summer of 1954, keeping with the tradition that the new model year begins in the fall of the previous calendar year. The 1955 Corvette and Chevrolet came with the new engine, standard in the Corvette and optional in the sedans and wagons. The factory ran official entries for Corvette in early events that year and the various magazines were duly impressed with both.

But habits die hard. It took months for the street racers to admit that the Chevy was doing to their Olds and Buick stockers what the overhead-valve V-8s had done to the flatheads. In terms of expectations, it wasn't until 1956 that engine tuners and amateur racers began taking a close look. What they saw was the future of the sport.

The hobby swap

Now that nostalgia has its hand deep in our pockets it's hard to credit what people did to the secondhand racing cars of previous eras.

In the late 1950s and early 1960s, they put Chevy engines in them.

At the time, it made sense. Ferrari and Maserati in particular always made sturdy cars, surely because the classic Italian races of the time and earlier were true road races. They had the Mille Miglia and the Targa Florio, on public roads closed for the occasions but not improved, and containing all the potholes and ridges and curbs and shoulders to which public pavement is heir. Thus, the basic sports model was strong.

Next came rapid evolution. As we can see from the specs and the photos, road racing cars during this time were becoming smaller and more powerful, faster and more specialized. The sports Ferraris especially went through the stage of being small and temperamental, then larger and just as fussy, then the same machinery packed into a smaller envelope. They were generally a season ahead of the rules, that is, the clubs learned what was being gotten away with and closed the loophole while the factories and tuners were seeking other ways around.

The third factor was that virtually every region (chapter) in the SCCA had at least one member with the means and enthusiasm to buy these racing cars. It worked best, by the way, if there were two or three such sporting gents; gave them someone to compete against.

What happened first, was that these sports would buy the best and newest from Italy, that is, Ferrari or Maserati. They had good men working on the cars and driving them and they contested for the win in local events and a good place in the nationals: Hansgen and Kimberly and Spear and Company had better cars and preparation, and probably skill too.

Every year came a better car, so the owners who got them new, got new models every season. Nostalgia not being here yet, they sold last year's car at bargain prices to racers who didn't have as much money.

The new owner could afford the car but he had trouble with the upkeep. A Maserati six or four, or a Ferrari twelve, six or four, isn't a simple engine. The pure racing engines come with extra parts and requirements: It would take 200 pages to detail how to assemble the valve gear of a Bristol six, or the built-up crankshaft of an early Porsche RS.

Thanks to Chevy and GM and the hot rod crowd, the second owner of the exotic car had an easy way out ... Chevy V-8, with Corvette parts and Corvette gearbox. The configurations were the same, some of the time, and the earlier racing cars especially had lots of space under the hood. You fabricate engine and transmission mounts, adapt the output flange from the gearbox to the front end of the driveshaft, work out the throttle and fuel line and hoses, and so on, rework the tunnel so the gearshift has a place to go and *presto!* another Ferrari-Chevy is ready to race.

There were such conversions by the score. Just as every region or track had its wealthy sportsman, so did every region or track have the team that bought the older car and converted it. The people who wrote the programs surely ran out of hyphens, so popular were the Maserati-Chevys, Ferrari-Chevys and Jaguar-Chevys.

How'd they do?

Not terribly well. The factors that made them practical worked against them on the track.

First again, they were last year's cars and the factories were learning at a great rate, so the hybrid was heavier, with inferior brakes and suspension. And just as obvious, with a few exceptions the drivers of the secondhand cars weren't as fast as the drivers of the new cars.

So they filled out the program and gave rides to new drivers and kept some good sports in a good sport, all of which surely proves that winning isn't everything, no matter what they say in football.

HWM-Chevrolet

The first effective application of Chevrolet's new small-block V-8 was, by definition, an engine swap.

Tom Carstens was an Allard pilot, following his career in midgets, who hailed from Tacoma. He was aghast when, in the movie *The Racers*, the plot called for the destruction of a lovely English car, an HWM. He was happy when his inquiries showed that the wreck was done with models and mirrors, that the movie company still had the car and that all he'd have to do was buy the HWM along with two Ferraris and two Maseratis the company wanted to get rid of.

Carstens flogged the others and began work on the HWM. It was a lot of work. The car began life as a convertible, you could say, with a small four-cylinder engine that could be converted to run on alcohol or gas, as a sports car or a formula car. It had a stiff (for its time at least) frame and what had become the classic modern suspension of transverse leaf spring and A-arms, independent front and rear.

Carstens' experience with his Cad-Allard stood him in good stead as he worked with hot rodders like pioneer Vic Edelbrock to develop and build the Chev engine. The engine was enlarged from 265 ci to just less than 300 ci, within the SCCA's class C limit. Compression ratio was raised to 10.5:1 and Edelbrock used his own camshaft and valvetrain kit and an intake manifold with three two-barrel carbs. Just about every part in the engine was reworked, balanced or polished. The stock engine was rated at 160 bhp, and showed 130 bhp on Edelbrock's dynamometer. (This might not have been a fudge from Chevrolet, as there are several methods of testing power.) No

Bill Pollack in the HWM-Chevrolet takes the lead at the start of the Pebble Beach race in 1956. The HWM was originally built with a smaller four-cylinder engine and with removable fenders and road gear, so it could be raced as a formula car. Tom Carstens adapted in a different direction, with the first effective use of Chevrolet V-8 in road racing. Road & Track

claim was made for Carstens' finished engine but it surely came close to being double the stock power.

The team adapted a Jaguar gearbox to the Chevy engine, needed because the timing was poor. By 1956, the Corvette design team knew its car needed four speeds, but the Corvette four-speed transmission wasn't offered until 1957. (By another quirk of timing, Chevrolet also released its fuel injection in 1957. The

Perhaps there's a lesson in suspension tuning as well as engines here: Bill Pollack in the HWM-Chevrolet leads a wheel-spinning Cadillac-Kurtis, driven by Lou Brero, out of a tight turn at the Seattle Seafair Airport in Washington. Road & Track

injection system was supposed to smooth and improve torque, to make up for the three-speed box, while the four-speed was to close gaps in the torque curve as shaped by the carbureted engine. Odd, how these things work out.)

The HWM got a Halibrand quick-change differential, a straight sprint car item adapted in this case for independently sprung and located halfshafts. The disc brakes also came from Halibrand, which was one of the few outside suppliers at the time. Jaguar's Le Mans win had proven the usefulness of disc brakes but most of the racing world was still crouched over the drawing board.

Carstens' driver was Bill Pollack, who never took prisoners and was the best choice for a car with more power than its 1950 chassis could use. The car was fast, and could win class and stay with the Ferraris off the corners, but it didn't have the top end or the cornering power to ensure consistent wins.

In other news, Torrey Pines—once an army base, soon to become a golf course—was the scene of its final sports car event, a six-hour enduro won by a C-Type Jaguar. Five months later a second final meeting was held, with a sprint race won by a 300S Maserati (with a big inline six—perhaps the best-sounding one ever built) and another enduro, taken by a D-Type Jaguar.

Briggs Cunningham sold his factory and formed a team of Jaguars, as the Le Mans organizers took a step toward rationalizing their event by changing the rules. Prototypes were limited to 2.5 liters, supercharged or not, and production cars had to be made in lots of 100 or more. To be legal a car had to have two doors, two seats and a windshield, and dimensions for all were carefully spelled out. The rules clearly excluded the low-production racing machine powered by an American—that is, big V-8—engine.

Another round was fought in the old war as we began to debate amateur versus professional drivers. The magazines were filled with discussion. Remember back in ARCA days, when the amateurs said no bad things about the professional dirt or oval racers? That was in the past. One way the new sports car fan proved his or her worth was carping or sneering about American drivers. It wasn't the money, as nobody who opposed American pros had any objection to Formula One drivers getting paid for their efforts. This was approaching the height of the Big Fin Era for domestic cars. By extension, those who didn't like big flashy cars also reviled the people who drove them and so on to Indy, stockers and the like.

The organizers and sanctioning bodies weren't much better, as the clubs fought each other for power. SCCA drivers who went out of the United States could race for money, but when the same promoter put on a race in New Jersey, the SCCA yanked the licenses of every member caught in the act.

On the bright side of sportsmanship, in May 1956 Pete Lovely offered for sale his famous Pooper, which was both national champion in its F Modified class and licensed for the street. (Surely that was the last time those two facts appeared in the same classified ad.)

Just as Le Mans rules didn't follow FIA standards, SCCA regulations did not conform to either one. Models eligible for production car racing had to be produced in lots of 500, for instance, so the D-Type Jaguars that qualified as production for Le Mans were modifieds in SCCA: good news for XK120 owners and Corvette racers, but another threat to the homemades.

Oh, yes, Cunningham bought four D-Type Jaguars for his team. Three crashed in one weekend, at Road America, but Cunningham shrugged and told his drivers they were there to race, not preserve.

Mercedes SLS by Porter

There was room for thought and for innovation and even for inspiration (sometimes defined as, "If you see a good idea, borrow it").

Chuck Porter owned a bodyshop in Hollywood, and one day he stumbled across a wrecked Mercedes 300SL. This was a rare and unique car, a sports coupe patterned after the make's world-shattering racers, winners of every long-distance event then held. Like Cunningham before him, Porter was a practical man and could see lots of uses for the parts still intact. He bought the wreck and hauled it to his shop.

Logically, he realized there wasn't much he could do to improve the car in terms of engineering. He straightened the frame—no easy task—and replaced all the broken bits and stripped off the steel coupe body. Sheets of thin aluminum were shaped by Jack Sutton, the best panel bender on the West Coast at the time, into a reasonable facsimile of the famed Mercedes lightweight racing roadster, the SLR.

Sports Car Illustrated had fun with the rumors about Porter's project. Seems some people thought he'd acquired one of the former team cars, the magazine said, adding that because the racing engine was an eight and the road engine a six, those who fell for the story were those who couldn't count above six.

The joke turns later, but for now, Porter's recreation was a virtually stock chassis, suspension, brakes and engine, covered with a lighter and lower and slicker body. The Mercedes six displaced 183 ci and with mild modification, according to *Sports Car Illustrated*, would produce 240 bhp. The car weighed 2,855 pounds, with fuel and driver. Not light by standards of the day.

Porter's car was reliable and fairly easy to drive. The road Mercedes used an odd rear suspension, a swing axle with low pivot point. It was less tricky than the lower-priced but classic Volkswagen Beetle system, but it wasn't good for racing.

Sports Car Illustrated ran the Porter car against a stock 300SL coupe and found the roadster many seconds per lap faster than the stock car, mostly

This replica Mercedes-Benz racer was built by Chuck Porter, who owned a bodyshop and a wrecked Mercedes coupe. It ran first with the original Mercedes-Benz six, but when the racing got tougher it got a Buick V-8. Road & Track

because it was easier and more predictable to drive. In its first races the Porter SLS—the second S stood for Scrapyard—was a class winner, nearly but not quite as fast as the three-liter Ferraris and so on.

There's a postscript here. Porter's SLS was written up in *Sports Car Illustrated* in November 1956. One year later, in November 1957, the magazine published a sequel saying "that the only good substitute for lots of rectangular money is lots of cubic inches."

You guessed. Porter swapped the Mercedes six for a Buick V-8. Seems he'd raced the stock Mercedes engine, then added a supercharger; there he was next to a production Corvette and they both put pedals to the metal... and the Corvette kept the lead.

The swap was aided by Max Balchowsky, who was *the* Buick man, a job he'd taken because nobody else applied. The Buick V-8 was different. It appeared in 1951, after Cad and Olds but before Pontiac and Chevrolet got their new engines, in keeping with GM's marketing plan. Buick's engineers had opted for efficiency through better mixture and flow, and they got that, or so they hoped, by using small valves, angled in the heads. The cylinder bores were 90 degrees to each other and 45 degrees from vertical, normal for a V-8, while the valves were straight up-and-down. These were called nail valves because the valve heads weren't much wider than the stems.

With a de-stroked crank to bring the engine within the SCCA's five-liter class B and with all the hot equipment Balchowsky could find, the Buick produced about 290 bhp. The builders had to move the engine back in the frame, the V-8 being shorter and wider than the inline six, and machined an adapter to mate the Buick to the Mercedes transmission. More power called for a larger radiator and bigger brakes, but when the job was done the Mercedes weighed 120 pounds less than it did with the stock engine.

Making a true story bearably short, it wasn't enough. By late 1957 or early 1958, the ante had been raised. A lot.

Barneson-Hagemann Chrysler

But first, more cubic inches. John Barneson was an Allard driver in the days when Chrysler's Hemi V-8 was the biggest. He wrecked the Allard in 1955, about the time the model wasn't at the top of the heap, and decided to take the next step. He commissioned a body and frame from Jack Hagemann, with power to come from a modified Hemi.

Hagemann was a top man and knew racing. The special got a wheelbase of 95 in.—balanced between quick turning, which comes from being short, and stable, which comes from being long. The frame was laid out around the engine, with the V-8 taking what's now known as a front-mid-engine configuration; that is, the engine was within the wheelbase ahead of the driver. Springs came from torsion bars and the front and rear axles were both live, in keeping with oval track practice and the smooth surfaces of western tracks. The alloy body was slick and pointed in front, with a hint here of Cunningham, there of C-Type Jaguars. The car had as much road equipment as the rules required, no more. It was compact, with the mandatory spare tire snuck beneath the fuel tank and the lights hidden in the grille.

The 375 ci Chrysler was claimed to crank out 370 bhp, using a hot cam and four two-barrel carburetors. Engine builder George Naruo said it would run all day at 6000 rpm, with a safe flash limit of 6500 rpm.

According to the men who talked to the builders, the Barneson-Hagemann Chrysler weighed 2,180 pounds dry. This may have been a target weight. It may have been wishful thinking. The Chrysler Hemi was huge and heavy, several hundred pounds heavier than even the other V-8s, so if Porter's Bu-Merc was heavier than Barneson's Chrysler, then one man had overbuilt or the other was hoping to scare the competition.

Barneson's car was a contender in class and for overall victory for the next couple of years, but it never overshadowed the field and that was surely because the mighty old engine was too big for its power output.

Ol' Yaller the First

Get out your handkerchiefs. Prepare to delicately dab your eyes or to wave with passion, depending on which side you choose... except that there really was only one side in this epic.

You simply had to root for Ol' Yaller.

Max Balchowsky was an interesting and entertaining man. Exotic name notwithstanding, he was born in West Virginia, was a gunner in a bomber during World War II and moved to California to help his brother repair cars, at which point he became a hot rodder.

He also became a sports car enthusiast and racer. Once again running against expectations, Balchowsky began his racing career with a stock Jaguar XK120. It wasn't a challenge, so he built up a 1932 Ford roadster, the classic hot rod of the day except that it was powered by a flathead Cadillac V-8. In those early days, the roadster was a contender, if not a series winner, especially when it got an overhead-valve Buick engine.

Next Balchowsky became the man to see about Buicks. At that point racing friend Morgensen left his homebuilt racer, big and ugly but often fast, in Balchowsky's shop. It was too good a chance to pass up, so Balchowsky "reworked everything except the steering wheel and the seat."

The Barneson Chrysler roadster had the huge engine set well back in the frame, for better weight distribution, and a beautifully shaped aluminum body. Road & Track

Enter here a factor the press doesn't often tell you about: History is dictated by the people who'll talk to the historians.

The press loves a good story, and Balchowsky has few equals when it comes to quips and yarns and making friends. At the time he began racing his revamped special, there was a Disney movie about an ungainly old hound with a heart of gold and nerves of steel: *Old Yeller*, or Ol' Yaller, as they said in the movie's dialect. So Balchowsky painted his racer yellow, brush marks visible ten feet away, and nicknamed it after the other mongrel. He referred to his creation as "that yellow pile of junk" and he reveled in listing all the junkyard parts.

In fact, the frame was as stiff and strong as any in the field, and the suspension was predictable and controlled, even if the components came from Pontiac, Jaguar and Morris Minor. The souped-up Buick displaced 400 ci. It delivered 300 bhp, same as the

Details of the front left corner of the Barneson car show double-caliper disc brakes, rack-and-pinion steering and a headlight, located out of the way. Road & Track

Everything in the Barneson car was sited for efficiency. The spare tire was the meat in a fuel tank sandwich; there was one tank above the tire and another below it. The body was three main panels—front, center and rear—each held in place with Dzus fasteners. Road & Track

Another stage in the saga: Eric Hauser in the former Morgensen Special, now called the Lion Cage in reference (one assumes) to the Birdcage Maseratis. We don't often see frame rails outside the body panels. Road & Track

Ferraris of half the displacement... but the Buick engine was lighter.

Balchowsky registered the car in Idaho, surely so he could get a plate that proclaimed Famous Potatoes. Wife Ina, herself an expert mechanic, often drove the creation to the races.

Balchowsky's partner in the venture was investment banker/amateur driver Eric Hauser. Balchowsky and Hauser had the money to buy what they needed. The car was as carefully constructed as any racer ever built; every part was the best for the job.

It was just that they made sure everybody knew who the underdog was.

Ferraris had ruled for several years and while it was thrilling to hear the ripping-silk shriek of twelve little cylinders revving high, it was more fun to hear

By 1957 the car had become Ol' Yaller, Balchowsky having reworked the front end, the side panels and the engine.

Two years later Ol' Yaller had still another nose and an egg-box-style cover for the carburetors. Those wide whitewalls were a trademark of the car: Balchowsky said he ran recaps to save money. Road & Track

Action! Eric Hauser and Ol' Yaller slide into the marbles at the first Riverside race, while Chuck Daigh in the Troutman-Barnes Ford keeps the inside. Headed down the hill are Bob Drake in the Aston Martin, and Bill Murphy in the Buick-Kurtis 500X. Road & Track

Ak Miller enjoyed the Mexican Road Race so much that he built a streamlined Kurtis with Chrysler V-8 and ran in the 1957 Mille Miglia, one of the last true road races. Here he leaves the starting ramp. Alas, the brakes failed and the car had to retire. Road & Track

the rumble of eight big pistons and watch the ungainly old collage thunder past the thoroughbred.

Even better was when Balchowsky told the press that he'd tried to buy an engineless Ferrari but the owner had raised the price, and to boot had offered to pay if Balchowsky would drive Ol' Yaller over a cliff.

Heck no, Balchowsky said in effect. So he built a new engine and the Ferrari guy hired a pro driver and darned if Ol' Yaller didn't meet and defeat the thoroughbred that very next weekend.

The crowd loved it.

And they kept on loving it. With Hauser at the wheel and later Balchowsky, Ol' Yaller won sixteen major races during its eight years of competition. During its best years, 1957 and 1958, Ol' Yaller and crew defeated Carroll Shelby and a 4.5 liter Maserati, Phil Hill and a three-liter Ferrari, Richie Ginther and a four-liter Ferrari, Chuck Daigh and the Troutman-Barnes Ford, Bill Pollack and the Murphy Kurtis 500X, and John Von Neumann and a Ferrari Testa Rossa.

In its first form, the car appeared in 1953. Its last win, with Hauser driving but the Buick replaced with a Chevy, came late in 1960. When, early in 1961, Hauser demolished the car in a spectacular 100 mph end-over-end crash and walked away to applause, it seemed only fitting that *Road & Track* published a lengthy obituary.

The junkyard dog had turned folk hero and humbler of the mighty. Nobody would be able to do that quite so well ever again.

Things were changing. The SCCA was getting more concerned with safety (not that Ol' Yaller wasn't safe; look at its record) and was getting tougher on

The skilled hands of Jack Hagemann again, this time creating a wedge-envelope for a Jaguar engine in a steel tube frame. Notice the Cozzi Jaguar's covered and legal headlights and the faired headrest. Road & Track

inspection. Even production cars had to have roll bars, a requirement the anti-Yank brigade said was "a hangover from the stock car circuit."

The SCCA had changed the names, from modified to sports, although those in the know would always refer to H Modified or C Modified. Production rules were changed again, to 150 examples, as an encouragement to the smaller makers.

Racing was becoming a professional business, as opposed to the charitable business exemplified by the air base and park venues. Riverside opened in southern California, and Laguna Seca appeared in the central section of that state. In each case, an entirely artificial racetrack replaced an adapted track: Riverside replaced March AFB and Laguna Seca replaced Pebble Beach.

Although no prize money was allowed in SCCA races, the SCCA didn't (mostly because it couldn't) try to enforce what private parties agreed to between themselves. A driver could be on the sponsor's payroll as service manager or development engineer or sales rep: obviously you could let your employees race, so how could somebody prove that you'd hired them to do that? It worked, in that suddenly a crop of American drivers (Dan Gurney, Richie Ginther, Carroll Shelby and so on) could run wheel to wheel with the world's best—and in Phil Hill's case could win the world championship.

Cozzi Jaguar

The spirit of road racing, and especially specials, was still amateur, in that guys were building what they believed in, or had on hand.

A good example of that was the Jaguar Special owned by Dan Cozzi. Cozzi bought a wrecked Jaguar coupe XK120M (M meaning a mildly souped-up engine), and threw away all the bent parts, plus body and frame. He built a steel tube frame, lighter than the factory's design, but kept the various suspension locations so he could simply attach the Jaguar front end, steering, rear end and engine, transmission and so forth. The body was by Jack Hagemann, whose work we've seen before.

Cozzi's roadster was especially clean, with inspiration from Ferrari as well as the racing C-Type and D-Type Jaguars. It looked more practical than, say, Ol' Yaller, except that the driver had no door. By 1957, the FIA and Le Mans had agreed on requiring two doors, and doors one could actually use. But the SCCA wasn't nearly as concerned with the world bodies as with keeping club members happy; which was only fair, because the SCCA was a member club, chartered to give the members what they wanted.

What they wanted was to race what they could buy or build, so while one *Road & Track* reader opined that he was "strongly in favor of anything that will get rid of some of the specials," John R. Bond said that the SCCA and the United States in general could darned well draw up their own rules if they so chose.

We did have our own rules, and they worked for us. Carroll Shelby was national champion, thanks to his 4.5 liter Maserati, in what was then called the unrestricted category. Most of the national SCCA races and the larger Cal Club races—the two organizations were still hated rivals—were won by Ferrari, Maserati or Jaguar, with an occasional win by Daigh in the Troutman-Barnes Ford, Balchowsky in Ol' Yaller or Barneson in the big Chrysler.

Two different Crosleys

In the smaller classes there were English and Italian production racers and there was room for self-taught diversity.

Consider first the Jones and de Camp Crosley special. The owner-builders ran a bodyshop, so they could shape and fabricate. The frame was truss type, meaning it had four main tubes running fore and aft, two per side and triangulated with smaller tubes. Rear axle was Crosley Hotshot, and front suspension was fabricated A-arms plus bits bought from Lotus, at the time a small English firm making club race machines. The car had Crosley disc brakes, Morris Minor rack-and-pinion steering, an MG transmission—Crosleys came only with three-speeds—and a gas tank from a French Panhard.

The engine was overbored a fraction, to bring the displacement to 748 cc, just within the class H limit

A beautifully finished Devin body on the Jones and de Camp Crosley showed how well the work could be done and how adaptable the fiberglass shell was. Road & Track

Beneath the Devin body of the Jones and de Camp Crosley was a sturdy frame, a combination of ladder and space, with aluminum interior panels and firewall. Road & Track

The Rattenbury Crosley looked like, um, it was efficient or something. What it really looked like was a car built by someone who didn't care what it looked like. Road & Track

of 750 cc. Then came a hot cam, bigger valves and a two-barrel Weber carburetor and tuned exhaust.

The body was a Devin fiberglass shell, beautifully detailed and painted. The builders were experts in bodywork, but they didn't see any reason to make for themselves what they could buy, and finish, for less time and money. They did some modifications, for instance the hood for the carb and smaller openings for the wheels.

The Devin was wonderfully adaptable. This Crosley was a small car, with a wheelbase of 82 in. and a tread of 46 in. front, 42 in. rear. It weighed only 880 pounds, but the Ermini from which the Devin was taken had such lovely lines and proportions that there's no sense of size, big or small. The builders spent $1,200 in cash, which in 1957 wasn't a small amount but was a fraction of the cost of a production racer, and they had a car that won class on the West Coast for several seasons.

The other Crosley special design was based on nearly the reverse set of values. The car was featured in *Road & Track*, in that magazine's series on sports car design. The builder was James Rattenbury, an amateur who liked nothing better than doing things his own way.

The Crosley was his third such project, he wrote, and it was becoming clear to him that if you wanted to beat the more expensive racers, you'd have to do it differently. So he built his own frame and located the engine directly amidships, set crossways to the direction of travel. This wasn't the first such application in

The obvious parts of the Rattenbury Crosley are the engine set sideways and off center in the middle of the car, the primary drive by means of a chain to the gearbox next to and aft of the engine, and the final drive chain and sprocket on the rear axle. Hidden by the inboard brake drums are the U-joints that let the rear wheels move up and down. The lateral location of the rear wheels is provided by the collars and leading arms, inboard of the rear hubs. This was a complicated and carefully designed car. Road & Track

history but it wasn't anything other people were doing at the time.

The gearbox went directly aft of the engine, driven by a duplex chain—what would in motorcycle terms be known as the primary drive. Outboard of the input and output sprockets was the clutch, borrowed from a Harley-Davidson Big Twin. The secondary or drive chain ran from the output sprocket back to an axle sprocket on—what else?—the rear axle.

Centering the mass was the phrase here. Later, when this configuration was more common, we were told that it gave the car a low polar moment of inertia; that is, all the weight was in the middle so the car wasn't so easily put into a spin.

What looks like one solid axle in fact mounts to the two saddles just inboard of the brake drums. Tucked away inside the drums are the U-joints for the independent rear suspension.

This was a complicated and detailed experiment. The car was compact, almost squat, and the driver was as close to the centerline as the passenger seat would allow. The body was shaped to wrap around the components with as small a frontal area as possible, rather than to slice through with streamlining. Again, the few concessions to road use were there because the rules said they had to be there.

The engine was stock bore and stroke, with stock compression ratio of 7.75:1. This was because Rattenbury was using a supercharger, a Roots type sized for an MG engine. Blower drive was from the crankshaft, by chain. Rattenbury wrote that the engine produced a steady 50 bhp at 5000 rpm and 12 psi boost, with intermittent peaks of 60 bhp at 7000 rpm and 13 psi. This was power at the rear wheels, and the builder said the output fluctuated because of tire slippage on the rollers of the chassis dynamometer.

Rattenbury did a lot of homework and burned a lot of midnight oil. When his car was described in *Road & Track*, he said that there had been some trouble, and various teething problems had slowed him down. It wasn't the first time that happened to a builder. The car didn't set any records, but one gathers from the tone of the article that the constructor enjoyed the chase more than the prize.

Crosley into Saab

Martin Tanner was an advertising man, based in Detroit and equipped with no technical training whatsoever when he became interested in sports cars. He raced an Austin-Healey but decided he'd have more fun if he built and designed his own car.

First, he laid out the basic design on his home drawing board: the wheelbase was dictated, or so he said, by the size of the board. Second, he made full-size cardboard components, so he knew what they'd be like before cutting any metal. Third, he bought a semi-

Jomars were brisk and rational, if not detailed to show standards. Agile chap at left is builder Ray Saidel. Road & Track

space frame another man had built to use Crosley and Fiat running gear. The frame was too heavy and didn't have the parts where Tanner thought they should be, but it was a starting point for his own frame, of steel tubing. The light tube sections were reinforced with magnesium panels for the floor, trunk, firewall and so forth.

The Crosley engine had the later, stronger cast-iron block with an Italian crankshaft; the Italians loved the Crosley engine as it was bigger than the baby Fiat, and they did amazing things with it. There was the usual hot cam and two motorcycle carburetors, tuned exhaust and Italian magnesium connecting rods. Claimed power at the flywheel was 45 bhp, or one horse per cubic inch, at 7000 rpm.

Front suspension was Fiat, with transverse leaf spring and brakes from the larger Fiat sedan of the day. Rear axle was Crosley, located by leading parallel links and a Panhard rod.

Tanner was a craftsman and spent weeks hammering out the aluminum body, panel by panel, and tacking each piece to the tubing framework. When all the pieces were in place, they were joined together and the body was shipped to a professional for finishing and painting. The car contained some odd little bits, like the Indian motorcycle headlights fitted with small 12 volt bulbs from a 1942 fire truck. How people find these parts is hard to imagine.

Tanner raced the car with the Crosley engine and did fairly well, then switched to a three-cylinder, two-stroke 750 cc engine from a then-rare Saab sedan and won a national championship in class.

Jomar

A parallel approach came from Ray Saidel of Manchester, New Hampshire. Saidel built a car that was like Tanner's in that it used a steel tubing frame and aluminum body, but the powerplant for the Jomar was English, a Coventry Climax with 75 bhp from 1100 cc. The Jomar weighed 1,100 pounds and Tanner's car weighed 720 pounds, so the power-to-weight ratio was about equal.

Saidel went with Volkswagen for the independent front and rear suspension, which indicated progress in that Volkswagen was very much in business in the United States and the Fiat 500 was in eclipse.

Jomar used a ladder frame, with independent rear suspension with upper and lower trailing arms, on top of which was also the torsion bar. Road & Track

The major difference was in the long run, though. While Tanner was a craftsman who savored his hobby, Saidel was a craftsman who reckoned to make his hobby pay: he allied himself with an English racing car maker and began producing race cars for other people.

Sadler Chevrolet

Evolution like that in Tanner's engine swap and in Saidel's emergence as a constructor for customers is a common theme with specials. Equally common, perhaps, is the idea that a person or team with scant resources can meet and defeat the big guys.

Enter Bill Sadler, and a car that reminds one of the chap who owned Abe Lincoln's ax: "Just like it was when Honest Abe split rails, 'cept of course it's had two new heads and three new handles."

Sadler was an Ontario craftsman who in 1954 dismantled a Jowett Javelin sedan (the Jowett was an English brand, now departed and not much missed). He built his own frame and used the Jowett rear axle and suspension, the 1500 cc flat four engine and the front suspension. He made a torpedo body, with cycle fenders, of aluminum. The car was competitive in class F.

Next, Sadler wanted more power so he installed a two-liter Triumph engine, with his own fuel injection. At about the same time, the alloy torpedo body was replaced by an envelope body, fully enclosed, of fiberglass. This was made at home, by using very small tubing to form the basic shape. Metal lath was attached to the tubing and then the lath was covered with plaster, worked and shaped and refined until it had become a male mold. Cloth and resin were laid on the plaster, and when that layer hardened it was thin enough to be finished and reinforced, and thick enough to serve as the body shell.

The slippery body and Triumph engine were doing well, until Sadler got a chance to buy a Chevy V-8 in 200 bhp Corvette trim. The engine was the same length as the four-cylinder Triumph, so the V-8 was adapted to the Triumph's gearbox, which proved to be reliable despite the doubling of power.

The extra power was too much for the old suspension, though, and wheelspin limited speed. Sadler reasoned that independent rear suspension would cure that, so he machined down the differential from a 1940 Ford. He cut off the axle housings and made flanges, so the new halfshafts could slide on the J-joints. The differential was mounted solidly in the frame, and the rear wheels were located by two long A-arms, with their pivots below the differential. The new suspension was sprung with a transverse leaf, as seen in old Fiats and later on the 1963 Corvette Sting Ray.

The new rear axle and suspension and then a preselector gearbox were so different that Sadler went for a new frame. He laid out a ladder-type frame, two huge 3½ in. longitudinal tubes, with lighter-gauge tubing used for attachment points for suspension and

This is a mid-life incarnation of the early Sadler Special. It should have the later frame—this is the streamlined fiberglass body—but the exhaust pipes imply that the photo was taken with the Triumph four rather than the Chevrolet V-8. William Green Motor Racing Library

drivetrain, and with a subsection of tubing to protect the driver-passenger compartment.

An additional consideration was the new FIA requirement that the inside of the cockpit be at least 47 in. wide. Sadler's second body was this wide. Furthermore, the original frame had tubes that intruded into the cockpit and he feared some technician might object, so the second frame was all clear across the midsection.

Sadler considered the FIA rules because he had an assignment in England, working for John Tojeiro, a production race car maker in England and the man who designed the AC Ace that became the Cobra. Sadler took his car along and raced in England. He did well with it on the club circuits there, although the local press wasn't impressed by the detailing of the car.

But it was Sadler's same old car, just as it was in the beginning except for the second frame, second body, second rear suspension and third engine.

We'll hear from Bill Sadler again.

Huffaker Chevrolet

Amateur originally meant somebody who does something for the sheer pleasure of it. When one considers the skill and time and expense involved in building a sports car body from scratch, one must decide that many builders really were amateurs.

An example in the other direction was Fred Knoop's Huffaker Chevrolet (Huffaker was an imported car parts firm). This was a normal special of 1957-58. It had a tube frame, an independent suspension and a Chevy V-8. The engine used fuel injection from Stu Hilborn, the man who devised a system that was both basic enough and effective enough to become standard fare on Indy cars. The Huffaker Special also used Halibrand knock-off wheels, another Indy staple and a good indication that the car's builders were willing to do whatever they needed to do.

The body was a Devin, as seen earlier on the Ermini and the Crosley and so on. It had been modified some, with air scoops and heat extractors and a generous air intake to feed the big radiator needed to cool an engine with the Chevy's power, but it was the same shell as used on smaller cars and even on rear-engine cars.

The Huffaker was a contender, as they say in boxing, as it had enough power to stay with anything

The Huffaker Special also used the helpful Devin body, modified here with larger air intake, secondary scoops for the front brakes and covers for the headlights. Knock-off cast wheels marked this special as one with few expenses spared. The network of small tubes above the headrest indicated that the SCCA was just getting serious about anti-roll bars. Road & Track

The Lister body was a mix of brutal curves and bulges and clever reading of the rules: By covering all the mechanicals with a giant bubble, the cowl and windscreen could be several useful inches lower than the other cars managed. The driver is a young Texan named Jim Hall. Road & Track

on the track. Knoop apparently drove with more spirit than finesse and race reports of the day have him in the lead, pleasing the crowd but not winning on account of spins or mechanical troubles.

Lister-Jaguar

American sports car fans took their cues from England, where the sport of road racing developed in much the same way; that is, postwar prosperity allowed more people to take part, improved machines made for better racing, and the best of the new blood brought original and even daring thinking into the design of the cars.

Brian Lister was one of the better English builders. As we've seen Americans do, Lister made his own car and it worked so well that he began building cars for customers. Listers first came with little engines, but when Jaguar's development program made that double-overhead-cam six live up to its promise, Lister came out with a sports car powered by a modified Jag engine.

Strictly speaking, such a car doesn't belong on these pages. Listers were sports racers, built to meet the FIA and national rules rather than for road use, but they were customer cars, production racers. They weren't homemade. Nevertheless, they were intriguing and successful cars.

Jaguar's C-Type and D-Type were done by engineers who had to prepare for all contingencies—sprint race today, Targa Florio tomorrow. They had to overbuild.

The club racing builders didn't. The Lister-Jag of 1958 used a tube frame, two big steel tubes, with wishbone front suspension and de Dion rear. It had huge disc brakes and an aluminum body, with even lighter magnesium panels available on special order.

The Lister-Jag had a wheelbase of 90.75 in.; tread of 52 in. front, 53.5 in. rear; and a claimed dry weight of 1,736 pounds. The souped-up Jaguar engine of the time gave more than 200 bhp, so with fuel and driver

Then there were the endurance racing rules, as at Le Mans where the car had to come with weather protection. On this early Lister-Jaguar, the patch of canvas spanning the fiberglass side panels is, so help the FIA, a top. Road & Track

the Lister could offer better than 1 hp for 10 pounds. Estimated top speed was 200 mph.

A clever trick was used here. The FIA decreed a windshield height, measured from the cowling that separated the engine compartment from the cockpit. Conventional cars of the day covered the engine with the body and had the windshield above that. Lister realized that if you formed a huge bulge, a boxlike shape enclosing the engine, you could lower the rest of the body—the cowling, in fact—and thus your car would be lower, with less frontal area, than the other cars in class.

And so it worked out. With nearly four liters of Jaguar engine, the Lister had the same frontal area as an 1100 cc sports racer. As a bonus, this muscular bulge, aggressive intake and oversized wheelwells—they had to be higher in proportion because the cowling was lower—gave the Lister a menacing aspect equal to that of a steam locomotive.

The Lister-Jag was an English Kurtis, one could say, or perhaps a latterday Allard, that make having retreated into selling speed equipment. It was fast and light and priced within reason—$7,700 with Jaguar engine, $4,900 with no engine. There was a cataloged option for a Chevrolet engine but none of the big teams who bought the Lister seems to have gone in that direction. One wonders why.

The Lister replaced C-Type and D-Type Jaguars as the vehicles for those using Jag power and for those who weren't going the big-motor Ferrari or Maserati route. For instance, Cunningham's team, starring specials graduate Walt Hansgen, was Lister equipped.

As still another side observation, note that there were good backyard builders who could field a competitive car for several thousand dollars, and those who spent ten times that, and those who could strike a balance, as with the Lister. On any given day, any of the approaches could and did pay off; the man with the most money wasn't always the fastest.

Scarab

Now we come to a tragedy and a brilliant success and a reincarnation and the peak of pure sportsmanship, all presented in several acts.

The first act begins with social history. We've seen that sports car racing was a sport a wealthy young man could play, like tennis and golf but unlike bowling or motorcycle racing. Road racing had become socially acceptable and public, with hundreds of thousands of people lined up at the fence, and with people like Cunningham featured in national magazines. So it came naturally that Lance Reventlow, heir to the Woolworth fortune, got hooked on cars.

The first Scarab, under test by Road & Track. *The driver is Lance Reventlow, with John R. Bond from* Road & Track *as timer-observer. The man with the tire gauge is Chuck Daigh.* Road & Track

An early shakedown run for the Scarab. This car was neatly done, with everything in its place and with engine and driver within the wheelbase. Road & Track

The Scarab had a Chevrolet engine in full racing trim, with fuel injection, many breathers and a magneto ignition at right front. The excellence and availability of the Chevrolet V-8 kept the domestic engine, and the home mechanic, in road racing. Road & Track

Reventlow was a handsome, personable man. He had boyish charm and millions of dollars and the easy appeal that comes from such a combination. There was no reason for him not to do things as thoroughly as they could be done. When he came of age—there was a bit of a flap over that, as he wasn't quite old enough when he began—he bought a stable of racing cars and earned his license and took his Cooper and Maserati racing in the United States and Europe.

Reventlow was quick and brave and a fast learner, so he went as well as his cars would allow. But he wanted more. At the close of the 1957 season, he decided to build an American sports car, to compete with the best from Europe.

Reventlow and his team manager, Warren Olson, laid out the basic format. It would be a conventional roadster, and it had to have an American engine. The Chevrolet V-8 was first choice.

Scarab Mark II	
Wheelbase	92 in.
Length	na
Tread	52 in. front, 50 in. rear
Weight	1,900 lb. dry
Engine	Chevrolet V-8
Displacement	339 ci
Claimed power	360 bhp
Top speed	169 mph
0-60 mph	4.2 sec.
Quarter-mile	12.2 sec. e.t.
60-0 mph	na

Make that only choice. By that time, only one domestic stock-block engine was in the running—Balchowsky's Buicks aside. There were some specialized specials with Chrysler V-8s and a handful of subsidized Fords, but the Chevrolet was the only

Up close the Scarab was a complicated machine. Looking backward from the middle of the car, you can see the solidly mounted differential, the inboard drum brakes and part of the de Dion tube at center left, behind the halfshaft. No expense or detail was spared. Road & Track

91

Late in the Scarab's career, by 1961, the car had developed and had been refined into a formidable competitor, easily a match for any of the big Britons or Italians. And the sponsor had arrived in racing. Road & Track

American engine that could compete on the basis of power per pound (if not per cubic inch). Without the Chevy there would have been no Scarab, nor would road racing have gone the way we'll watch it go. (It's also puzzling and entertaining to wonder how Cunningham would have done at Le Mans if the Chevy engine had been available when his enthusiasm was at its peak.)

Meanwhile, Reventlow began doing things right by forming a company and hiring a design and development team. He commissioned Ken Miles, Dick Troutman, Tom Barnes, Chuck Daigh and Emil Deidt, the best in their fields.

The Scarab, which was mysteriously named for a sacred Egyptian dung beetle, had a space frame of steel tubing—1.75 in. diameter for the main sections, 1 in. and 0.75 in. diameter for secondary parts. The front suspension was upper and lower wishbones with coil springs, and the rear suspension was independent, by means of de Dion tube and trailing arms, with coil springs.

The original car had adjustable rear-suspension geometry. The hub carriers had two machined plates, each eccentric. The two halves pressed together and were splined, so toe-in and camber could be adjusted for each wheel simply by the position the two halves were in when bolted together. This was a complicated and expensive piece of machining; after tests on the first car showed the optimum setting, later cars had solid hub carriers and fixed alignment.

The Scarab designers were among the first to realize that soft springs and lots of wheel travel kept the wheels on the ground, delivering cornering grip and forward motion better than stiff springs and minimum wheel travel. So the Scarab had generous wheel travel with the biggest shock absorbers on the market. Brakes were aluminum drum, dual leading shoe and something of a surprise in that the case for disc brakes would seem to have been made several years earlier.

The engine was in a mild (for racing) stage of tune. It was slightly overbored from the 1958 specification of 283 ci, to 302 ci, just under the five-liter limit for class C. The car had a mild cam, a high compression ratio of 10.5:1 and Hilborn fuel injection, which flowed more air and had better control than Chevrolet's injection system did although it wasn't as good for street use (no problem in this application).

The first car was built in five months. Early in the project's life the FIA dealt a severe blow in the form of a displacement limit: As of 1958, all engines would be restricted to a displacement of less than three liters. Reventlow wasn't daunted, at least not outwardly, and announced that they'd get to work on an Offy-powered car for international events. This didn't work out, but that comes later.

The Scarab was a beautiful piece of work. Each detail was exquisitely carried out. Its lines matched those of the Ferrari roadsters—no mean feat—and were slick enough to give the car every mph the engine would deliver. The first car was built at an announced cost of $25,000. Obviously this didn't include salaries and rent and such.

In the Scarab's first race Reventlow led, charged off course and came back to finish third, with a sick engine. He took the car back East, to Danville, Virginia, and was solidly trounced by the Lister-Jags of Walt Hansgen and Ed Crawford. In the second race of that weekend the Cunningham team piled up a thirty-second lead, until told to ease off. According to *Road*

& *Track*, Reventlow "never had a chance against the two more experienced drivers."

Back in the West, Reventlow led the Cal Club races at Santa Barbara, California, then spun out but came back to win, eight seconds ahead of Richie Ginther (who went on to Grand Prix status) in a three-liter Ferrari. Quoting *Road & Track* again, "The triumph... over Europe's best was highly applauded, although to consider Reventlow's car a backyard bomb is definitely stretching things a bit."

Perhaps.

Even so...

For the SCCA national championship race late in 1958, Team Flat Kat, as Reventlow called his opera-

Chuck Daigh, left, and Lance Reventlow jointly accept the Nassau Trophy from Lady Oakes, late in 1958. Road & Track

tion, had two Scarabs ready. Reventlow and Daigh were the drivers, versus Hansgen in the Lister-Jag and Sadler in the Sadler Chevrolet.

The contest was over by the end of the first lap, as Reventlow gunned through the last turn with inside wheels in the dirt, showered Hansgen with rocks and powered into the distance. Daigh was right behind, until the rear axle gave out. By the end of the race Reventlow had lapped every car in the field except Hansgen's. Reventlow "did not try to hide his pleasure," said *Road & Track*, "but at the same time there was no trace of gloating."

Owing to the odd nature of the SCCA points system then, and to the Scarab's teething troubles, Hansgen retained his national class C title for the year, but everybody knew who had the best car in class, and the Chevrolet-powered option for Listers suddenly became worth looking into.

What would the Scarab do? No fools, the *Road & Track* staff arranged for a full track test of the car, with Reventlow doing the driving.

Road & Track came up with some impressive figures: "Not unexpectedly, the car's performance exceeds any and all previous data recorded by us, or by anyone else for that matter. It is possible that the 5-liter Ferrari or 4.5 Maserati V-8 might improve on the figures here, but it should not be overlooked that the Reventlow crew has produced a much better handling chassis, which is no mean achievement."

The Scarab had changed some in the year of development, as the engine grew in displacement and in power. The crew told *Road & Track* that the engine in the car, which was the second Scarab built, was an ordinary engine, and that their better units cranked out 385 bhp and could go to 400 bhp in brief spurts, if the compression ratio was raised and the engine was revved to 7000 rpm. But because the milder engine had equal power to the competition and the chassis handled better, they wisely kept stress to the practical minimum.

Road & Track got a chuckle out of Warren Olson's confession that much of what they'd applied to the Chevy engine had come from the work done by Clay Smith, legendary Ford tuner. Because the secret to making any engine produce more power is to make it breathe better, it follows that those who could make the cumbersome, tortuously routed flathead Ford work would have no trouble with a design that started out well.

Road & Track writers made some other notes. For one, they were surprised at the double-leading shoe drum brakes. However, they noted that because the brakes were small, they warmed up quickly and were less liable to the grabbing and overperforming that the type is famous for. Plus, with driver in place and tank topped off, the Scarab had a full sixty percent of its weight on the rear, driving wheels. That helped traction, and because of weight transfer while braking, under full brake power the balance would be close to fifty-fifty, so each brake was doing the same work and they were better balanced.

Even so, with what sounds like extra weight in the rear, the Scarab's primary handling characteristic was understeer, with the front wheels sliding wide on turns. That was good in that the car could be balanced with power, just like the classic dirt or sprint car, but it's not what was expected. By the time *Road & Track* tested the car the suspension had been adjusted, with stiffer springs and a roll bar in front; the team could have taken the big rear tires off and used the same size as they had in front, but that could have cost them traction. No member of the magazine staff drove the car, but from the observer's perch the Scarab cornered with slight understeer, negotiable on demand.

As the Scarab sat early in 1959, it was the best special out there. Phil Remington, a pioneer hot rodder who helped make the Scarab work and went on to give equal help to Carroll Shelby and Dan Gurney, says that Reventlow matured into a top driver. Not world class perhaps, but as good as the car and the other men competing for national honors. If Reventlow had a flaw, Remington says, it was that the owner-driver wasn't a mechanic. He didn't really understand the engineering principles behind the design, or the choice of components or even of configuration. This was to become a shortcoming later. But then, with the first Reventlow project a sizzling success, the plan was to keep winning (which the cars did for the next several years) and to move ahead, to tackle even more challenging venues.

Meanwhile, with the Chevy engine equal to the big and more pure racing engines from elsewhere, the sport of road racing was about to undergo a monumental, perhaps terminal, change of its own.

Chapter 5

The Professionals 1959-1964

Rather than hang separately, the clubs band together, while the racers take the money and run

Late in 1958 the relentlessly logical Ken Miles penned an essay in which he asked, in so many words, why hang separately? His reference was to that old bugaboo, money. He was concerned with the issue of professionalism—that is, prize money for drivers—and he was even more concerned about the lack of concerted effort from and among the various clubs then involved in road racing.

The issue had been building for years. The forces of collision had been picking up speed for just as long.

On the one side, we had the pure sports, those who loved speed and daring for their own sakes. They re-invented road racing in the thirties, when professional racing was at low ebb, especially when it came to sport. To be frank, promoters stole and racers took dives, agreeing beforehand on who'd win and by how much. Not always, not much at the top, but enough to give an unsavory air to the, um, sport.

Against that, by the fifties sports car racing or road racing was a genuine business and a big business. Hundreds of thousands of fans turned out for the major events. The names at the top were household names. Millions of dollars were taken in by the promoters and clubs and track owners, while other millions

Truly a contest of speed, it's Charley Kolb in the Sadler Chevrolet and Fred Gamble in Ol' Yaller IV, with Jim Hall in his Chaparral looking to get between them, at Nassau. Road racing had become professional entertainment and there were those who thought the drivers deserved a share of the profits since they surely took most of the risks. Road & Track

95

Lime Rock, Connecticut, 1958, and the very first professional road race in modern America. Marshall Teague in a D-Type Jaguar with Chevrolet power leads Alan Markleson in a Ferrari Testa Rossa. Teague was one of the first truly professional stock car drivers and Markleson was a well-to-do enthusiast—and if it hadn't been for the pro series, they'd never have met. Road & Track

were spent by the racers and builders and, in some cases, sponsors (they simply had to be circumspect, and not paint names on the cars).

Then came politics. The racing division of the AAA, which always preferred single-seat racing on oval tracks and didn't like stock or sports cars anyway, had been taken out of business by the Le Mans tragedy of 1955. The car owners and promoters and track owners got together and agreed on a new sanctioning body, the United States Auto Club (USAC). There were also the National Hot Rod Association and the National Association for Stock Car Auto Racing (NASCAR), as well as the Sports Car Club of America and Cal Club.

The USAC and NASCAR were professional clubs that cheerfully raided rival turf, as in NASCAR drag racing and USAC stock cars, while experimenting with road racing and sports car classes. Early in 1958, for instance, USAC announced the creation of a professional road racing division. This made good business sense: the professional clubs could see no reason not to pay prize money and surely not to pass up the profits the SCCA and its various charities were raking in.

Ken Miles' approach was sort of enlightened professionalism. He asked how having races with the top drivers from SCCA, Indy, the stock car races and even Europe could have anything but a good effect on road racing. Nobody could refute that one. Most American amateurs had been raised, after all, on the notion that the Europeans were best, so getting Europeans over to the States was bound to improve the sport. Equally, while there were good local talents who didn't need outside income and talents who could use their Sunday expertise to help their careers on Monday, there were also talents who couldn't afford to race on their own time never mind run a Porsche Spyder or Lister-Jaguar.

Except that the SCCA, especially at the top, had a clear idea of sport for the sake of sport, and its explicit reason for being was to provide sport for the members rather than a living for drivers or even funds for good causes.

Beneath all this discussion about ideas was a legal battle in which the operators of one club would

Ak Miller in a Devin-Chevrolet he built for sprint racing, bracketed by Jack Graham in an Aston Martin and Indy driver Jack Flaherty with a Lister-Jaguar, at Laguna Seca, 1959. Miller raced on a budget, for fun, and said if you won a few bucks, that didn't hurt, either. Road & Track

yank the licenses of anybody caught driving for any other club. It happened in oval track racing and in sedan racing and on the sports car circuit.

But the money and the chance to learn, to race against the best, proved impossible to resist. Late in 1958 the Road Racing Division of the USAC presented its first professional program, at Lime Rock Park, Connecticut.

The first was . . . the first. It attracted a record crowd but was plagued by rain and a small entry of sixteen cars for one 100 lap race, in contrast to the scores of cars at SCCA races. The field included an Aston Martin, a Porsche and a long list of chassis with Chevy power such as Jaguars and Maseratis. Entries were varied, with stock car star Marshall Teague in the Jag-Chevy, and Indy star Lloyd Ruby in a Maserati-Chevy. Bob Said, an American who went to Europe to race and learned well, was in the Sadler Chevrolet, running as the Nissonger Special. The Sadler and the two engine swaps broke, and the Aston won.

Then things looked up.

"Sports car racing, at long last, comes of age in the west," proclaimed *Road & Track*.

The magazine's proclamation was justified. Everything came together and made the big time at Riverside. At that period, the track was a great location, close to freeways and population centers in Southern California, yet far enough away for the neighbors not to mind. The first professional race was sponsored by the powerful *Los Angeles Times*, which got the word out. And the race was sanctioned by the FIA and the USAC, and put on by Cal Club.

Because conditions were good, the field was great, from both the professional and the amateur ranks.

Phil Hill, Masten Gregory and Carroll Shelby, who were driving Formula One in Europe, came home and brought France's Jean Behra and Sweden's Jo Bonnier with them. Lance Reventlow and Chuck Daigh were locals, along with Dan Gurney in the 4.9 liter Ferrari and Richie Ginther in a three-liter Ferrari. Then there were Ak Miller and Ray Crawford (the underdog winner of the Mexican road race a few years earlier), Jack Rathmann and Bobby Unser from the Indy crowd . . . everybody who was everybody.

Daigh won for Scarab, after a race-long duel with Gurney and the thundering big Ferrari, in front of a capacity crowd of 70,000 people who all went to work Monday and told those who weren't there how much they'd missed (and if they didn't, the *Times* did).

Well now. We are embarking on a new era.

The prize money wasn't all that much. It couldn't come close to paying for, say, Gurney's Ferrari or Daigh's Scarab. Nor would the investments made by those makers return with interest in the form of sales. Chevrolet could afford to use racing as a marketing tool, but Scarab could race only for the sake of racing, no matter what it told the tax collector.

In addition, there's clear evidence that the locals were underestimated, by both the imports and the pros. The Europeans came over to drive good production sports cars; Behra had a brand-new Porsche RSK, which *could* have gained in deftness what it lacked in sheer power, but it didn't. The oval crowders harkened back to the days of tea-bag jokes and had to learn the hard way that stuffing a homemade V-8 into somebody else's stock chassis wasn't enough for the job.

At the top, the Scarab was as good a racing sports car as there was in the world, for the races in which it would compete. And Daigh was as good a driver.

That's the professional side.

There was also the human side. Ak Miller was sort of a professional-amateur, a man who'd tackle any project if the odds were against him. In that first pro race at Riverside, he drove his Kurtis-Chrysler, which was an outrageous project built for racing in Europe. He'd also built another car, a Devin-Chevy, which he loaned to Bobby Unser. The Devin had the better chance of winning, Miller recalled, and he knew Unser was the better driver, so he drove his old car until officials called him in for an oil leak they never did find, and Unser flogged the Chevy until the stroker crank let go.

Not that they cared all that much. Miller: "In those days we bought recapped tires for $20, and paid an entry of $20. For a total of $100 [not counting the car, of course], you could have a weekend of racing."

Miller had a partner who owned a sheet metal business. Miller supplied the car and driver; the partner paid half the bills in exchange for being able to buy and hand out pit passes for pals and customers. One weekend, Miller said, their team had 103 paid guests.

Could their homemade car and part-time help win the Main Event? Nope. "But we were competitive for fourth or fifth place, and it was fun. That was the main thing."

Devin Super Shillelagh

It was a time for great expectations. *Sports Car Illustrated* estimated that the Riverside pro race attracted a crowd that spent $300,000 to watch cars worth a collective $1 million compete for $14,000 in purse money. There was no balance. Instead, there

Devin SS	
Wheelbase	92 in.
Length	na
Tread	52 in. front, 55 in. rear
Weight	2,170 lb. curb
Engine	Chevrolet V-8
Displacement	283 ci
Claimed power	220 bhp
Top speed	131 mph
0-60 mph	5.7 sec.
Quarter-mile	14 sec. e.t.
60-0 mph	na

The Devin Super Shillelagh was a professionally finished Devin body with a fully upholstered and equipped cockpit, a full-width windscreen and all road equipment. Road & Track

was lots of enthusiasm and faith that road racing and sports cars were the coming things.

Perhaps as part of this spirit, Bill Devin, the man behind the all-purpose fiberglass body, took the next step and went into the sports car business. It was the sort of complicated transaction that would bring grief to other entrepreneurs in a later time, but basically Devin worked a tax deal and arranged to have a chassis made in Ireland. The semi-completed rollers would be shipped to Devin's plant in California, for engine and body and finishing.

The plan was simple. Use the Chevy engine as developed for the Corvette. Do a professional finish on the standard fiberglass body, which in 1959 was still timely and slick and aerodynamic. Provide a workmanlike frame and suspension. Keep the car simpler and thus lighter and less expensive than the sort-of-sibling Corvette. And you have a Corvette beater with Chevrolet parts!

And that's mostly how it worked out, at first and on paper.

The Devin SS (which began as the Super Sport but was changed to Super Shillelagh to honor the Irish connection) was based on Devin's racing experience, which was considerable. It had a sturdy ladder frame, independent front suspension with upper and lower arms, and independent rear suspension with de Dion tube and with the differential firmly fixed in the frame. Chevrolet offered the V-8 in 220 bhp and 290 bhp forms, and Devin went with the milder version because the car was light and thus didn't need the fussy extra power. The car had headlights, taillights and so forth in the legal positions, and optional spring steel semibumpers. It also offered full weather equipment and a trunk with space for two standard suitcases.

The Devin SS was positioned between the Scarab and the Corvette. It was quicker and more expensive than the mass-produced sports car as well as less aligned toward comfort and ease, while at $6,000 list

The Devin SS looked as lovely bare as it did bodied. The ladder frame used stout steel tubing for rails and crossmembers. Engine and gearbox were stock Corvette, suspension was fully independent and brakes were disc. Chassis was made in Ireland and shipped to the United States for engine and body. Road & Track

Bill Sadler in the Mark III—also known, when the rules allowed, as the Nissonger KLG Special. The car and name were the next steps: the car had a stronger space frame and more powerful engine, and Nissonger Corporation bankrolled the project. Often the rules required no names, as seen here at Cumberland, Maryland, in 1958. Road & Track

it was a lot less expensive than the Scarab, which could only win races.

The Devin was a good car, impressive for its time and circumstances. *Road & Track* tested an early example and its drivers enjoyed the car. Plenty of power and traction, they said, with high grip and mild oversteer created by properly balancing the front and rear suspensions. The Devin was much better on the open road and in traffic than they'd expected the kit-descended car to be, although the ride was rougher than a sedan graduate would enjoy.

Which was all well and good. The problem came later. The Devin SS simply didn't sell. The plan was to produce 150 examples, but actual numbers never came close to that. Devin had bitten off more than he could swallow. Building cars took more time and money than expected, there was no dealer organization so distribution had to be farmed out and more people talked pure sports car than wanted to pay for one.

A side issue might have come into play here, in that the SCCA required several hundred cars to have been built before a model could qualify for production class racing. The Devin couldn't qualify, so it had to run against the incredible Scarab, the Lister-Jaguars, the Ferraris and company. Maybe it could have done well if prepared for racing, but Bill Devin had all he could to produce the SS and the less expensive and smaller Volkswagen Devin introduced at about the same time. There was no time or money for a factory racing team and no privateers took up the challenge.

The Devin SS went off the market. Too bad, in many ways. For generations the English had a tradition of the small maker using mass-produced components and they did well by it with the Lotus and Cooper. Americans, though, have never been as willing to buy the odd brand. And since passage of the infamous federal safety act of 1966, there's been no legal way for the small manufacturer to certify a car for public sale, so we've probably seen the last of that breed.

Sadler Mark 3

There was no shortage of ambition, though.

Late in 1958, not long after the early professional races were run, the SCCA recognized the obvious and took that first important step: SCCA members were now allowed to compete in races put on by other organizations, such as Cal Club or USAC or whoever; they just couldn't collect any prize money they won. (Cynics might wonder if the threat of lawsuits alleging restraint of trade, as seen in other sports, might have had something to do with this.)

The door was open a crack. Thus, when Bill Sadler won the 1958 Watkins Glen race, following a national in which he led the Scarabs and Listers only to break, his work attracted the attention of a sponsor, Earl Nissonger, who owned a major importing firm.

Nissonger backed the later races of Sadler's Mark 2 Chevrolet, and he bankrolled design and construction of a completely new car, the Mark 3, or Nissonger

KLG Special. The Mark 3 used Chevy power, of course, but bored and stroked to 327 ci (a moderate increase under the circumstances) and with 11:1 compression ratio, Hilborn fuel injection and a claimed output of 340 bhp.

The frame was space style, with smaller square section tubing and lots of it. If you use small tubing properly braced and have the tubes far apart, you get the same resistance to flex that you get with bigger tubes close together, but lighter. That's what Sadler did.

The front suspension was upper and lower arms with angled coil springs over shocks, and rack-and-pinion steering. The rear suspension began with a solidly mounted Halibrand quick-change differential, using bearing carriers and hubs and swing axle, as in independent halfshafts, with the rear wheels located by trailing arms and radius rods. The gearbox was Chevrolet, as used by virtually all special builders after the factory did them the favor of going to a four-speed in 1957.

Sadler took a page from the Lister design book by making the body sweep up from the radiator intake, over the top of the engine and then down again to the firewall and cowling and windshield. The windshield of the Mark 3 was as high as the hood of the Mark 2, which proves that point.

Sadler had the time and expertise—he'd spent a year working in England with top racing car designers—to try all the ideas that sprang to mind. For example, he made detailed adjustment to the suspension and alignments, and had the engine and gearbox offset to the left to give the driver more room and to rebalance the driver's side-to-side weight. This last is odd. Road races are traditionally run clockwise, with more turns to the right than the left. Thus, because more weight on the inside increases cornering power, the driver usually sits on the right. One would think the engine would be plunked in the center, to utilize the rightward bias, but Sadler didn't agree, and he may have been correct.

The Mark 3 was a small car, with a wheelbase of 86 in. and a tread of 53.5 in. front, 51 in. rear. The car wasn't track tested by any publication, but *Sports Car Illustrated* did a detailed analysis and said that according to the weights of all the parts calculated during assembly, the car itself weighed 1,650 pounds dry. The oil tank held twelve quarts and fuel capacity was 47.5 gallons, so the Mark 3 probably weighed a bit less than 2,000 pounds on the starting line. It was a competitive car, able to run in front with the Scarabs, Ol' Yallers, Listers and Astons, but perhaps because it wasn't quite as sophisticated or driven by men who had total ambition, it never won the major races. A journeyman effort, somewhere between the top contenders and the for-fun entries.

Bocar

There was a lot of enthusiasm going around. In a short time the normal backyard builder had gained access to technology, thanks to Chevrolet and to the refinement of fiberglass, which had previously been the province of the top factories. New makers with big plans came from all directions.

One new car was Bocar, named in abbreviation of builder-designer-owner-sponsor Bob Carnes, an aircraft engineer from Denver. Carnes was a good talker, perhaps too good. He attracted the attention of the

Ak Miller, left, with his Devin-Ford and the Bocar XP-5, right, at Salt Lake City, 1960. Miller later said he was pleased to win this race mostly because the Bocar driver had been rude before the race began. Road & Track

crew at *Sports Car Illustrated* early in 1959 and was invited to bring his roadster out to California, which he did.

Sports Car Illustrated had a grand time. The Bocar, designated XP-5, was conventional in that it used a space frame of chromoly tubing; a fiberglass body; and a stock, 290 bhp, fuel-injected Corvette engine with four-speed transmission. The engine was mounted in the front, driving the rear wheels. The front suspension was dual trailing arms and torsion bars, like Volkswagen's or Porsche's, and the live rear axle was located by trailing arms. The rear axle came from a Chevrolet, and the drum brakes from Buick, which was promoting its aluminum-finned drums.

Carnes must have been a charming man, or the *Sports Car Illustrated* test crew was ready to be impressed. Using a 3.70:1 final drive ratio, the actual car was said to have hit 150 mph at Riverside, a declaration based on watching the tachometer. The engine was an older unit, having raced at Pikes Peak. The magazine testers said they limited revs to 7000 and that was 150 mph, but with a taller gear, which a buyer could order, the same engine speed in top would be 160 mph. So the Bocar was officially declared a 160 mph sports car, in the headline no less.

This isn't to say the car *couldn't* have done such a speed, only that it didn't go that fast when under the clock. But the headline stuck in the minds of readers, and for years afterward people with good memories would talk about the Bocar and how fast it was.

Judging by the timed figures in the test, the Bocar was as fast as a 2,000 pound car with 290 bhp at the flywheel would have been under those conditions. It was right along there with the Devin SS, with no hope of catching the Scarab.

Carnes offset his engine slightly to the right, to balance the car's bias to the left when the driver was aboard. This had another benefit: a wheel had been improperly installed and fell off during the track test, but the car rolled along on three wheels until coming to a safe stop, so far from the wheel that it was nearly lost.

As an inadvertent hint of things to come, the engine was well back in the wheelbase and static weight distribution was forty-four percent front, fifty-six percent rear. The tires were the same size front and rear, 6.00x16, which was a traction limit. But the power of a car in this class was such that rearward bias was useful, as we'll see when the engines begin to relocate.

Carnes told *Sports Car Illustrated* that he'd be building more cars for customers, with choice of Chevrolet or Pontiac engine and various gearing and gearbox options. Just as with the Devin, though, only a handful ever came from the shop and the XP-5 never won anything of note. Ak Miller recalled beating one in a race in Utah a few years after this, when he had a Devin with 427 Ford V-8 and the Bocar owner made the mistake of bragging before the race that Miller was sure to lose. Purse money and the glimmer

Bocar XP-5	
Wheelbase	90 in.
Length	na
Tread	54.5 in. front, 52 in. rear
Weight	1,650 lb. dry
Engine	Corvette V-8
Displacement	283 ci
Claimed power	290 bhp
Top speed	150 mph
0–60 mph	6.0 sec.
Quarter-mile	14 sec. e.t.
60–0 mph	na

of sponsorship brought out many builders and hopefuls, and winning was tougher than it looked from the sidelines or had been only a few years earlier.

Kellison GT

Another parallel came from the availability of fiberglass kits. Devin was one of a score of men who learned the technology and measured the interest and embarked on the business of supplying parts for the hopeful.

Devin's interest was mostly racing. More commonly, the kit company made body shells and parts to adapt said shells to whatever chassis was handy. Do-it-yourself was becoming a watchword about this time and hot rodding was a major youth activity, and there's always Yankee ingenuity.

Jim Kellison was another major force in this market, working mostly with street sports car bodies, coupes or roadsters. Kellison was the third leg of a

Bocar XP-5 under test by Motor Trend in 1959. The crew liked the car, which got plenty of ink but didn't achieve the production success its builder had hoped for. Nor did it win races. Peterson Publishing Company

The menacing Kellison GT. Like other firms in the field, Kellison began with fiberglass bodies and kits, then tried to expand into limited production. The Kellison was a couple notches too rough-and-ready for the GT market, though, and too big and soft for the racetrack. Petersen Publishing Company

triangle: Devin did a thriving business in racing bodies and couldn't quite make the leap into producing complete cars; Carnes was technically adept but lacked the resources to compete on road or track; and Kellison was firmly established in the kit field but didn't make cars.

Motor Trend spent a few pages on Kellison in 1959. The object was a coupe and the purpose seems to have been an attempt to move Kellison into the sports car bracket from kits. Kellison supplied a demonstrator, a coupe body with full road equipment and a 1956 Corvette V-8. The frame, otherwise undescribed, was one of Kellison's own.

The article wasn't quite a test; the writers said the engine was tired so they didn't run the clocks; if the car wasn't up to snuff, why drive it? In the same manner they had a top driver along, the man who was to race the car the next weekend and he said he couldn't win because the engine was weak; same question. Nor do we get to learn the wheelbase or weight or brakes or suspension or anything except that they blew a tire. In short, an odd excursion.

Perhaps Kellison learned something from the response to this article, because he was in business for years after that, selling a full line of nicely done fiberglass bodies along with frames and parts and plans. However, the completion rate—the percentage of kits sold that actually became registered, usable cars—was only about one in ten. (Another kit maker confessed that he made most of his money selling the catalog and plans, because what most people wanted was the dream.)

A short, unhappy life

That's part of a title, and the title ran with one of the subject's best true confessions.

In an article for *Sports Car Illustrated* in November 1959, Charles Beaumont detailed his adventures as a specials builder. He began with the poignant statement that the average fan spent the average races draped over the fence, muttering, "If they'd only let me out there, I'd show the bastards."

So, Beaumont wrote, he and his pals pooled their resources and began building a special. They estimated $2,000 and six weeks. They learned the hard way that building a special at home, with your pals, is a grand way to use up time and money and often friendship as well. The slogan became, Got to get the bugs out, until as Beaumont stated, "It was no use. We'd never get the bugs out, never, and if we did new ones would come to take their place."

Beaumont went on to buy a production car.

This writer ran a souped-up MG as a modified for one season and learned the same lesson... the lesson being building a special was much more time and work and commitment—we should be committed for trying it, the joke used to run—than the magazines or the catalogs ever talked about.

Instead, another success story.

Ol' Yaller II

When last seen, Max Balchowsky's Ol' Yaller, the best and homeliest special on the West Coast, had been winning and was slightly behind the times.

Balchowsky knew it. He and Hauser parted pals, with Hauser taking the car, which he'd supplied in the first place, and Balchowsky retaining the Buick engine, which he put to nonrace use.

Balchowsky retired the engine because Buick had come out with a new, larger engine, displacing an astonishing 401 ci. By using mild cam timing and compression ratio, 9.5:1, and by not winding the engine beyond 5700 rpm even in the heat of battle, Balchowsky got an honest, unstressed 305 bhp with six carbs on a log manifold. His crew tried Hilborn fuel injection and gained 24 bhp, but Balchowsky thought they already had all the power they could put on the ground, so he saved his money.

The new engine went into an entirely new car. The frame was a mix of ladder and space, with four parallel main tubes with cross-members, weighing in at 110 pounds. Balchowsky recalled that a member of a certain high-buck team with a special of its own—surely Reventlow's although Balchowsky won't name names—said the frame was too weak. Balchowsky proved him wrong by running into barriers—accidentally, of course—and emerging unhurt.

The front suspension was Pontiac A-arms and Morris Minor steering. This wasn't a matter of using junkyard parts, as the pieces were cut apart and rewelded and x-rayed to be sure they were strong. Balchowsky didn't waste money, and the Pontiac and Morris components were what he knew would work.

A typical Balchowsky device was that the front suspension arms were liberally drilled. Not for lightness; there wasn't enough metal missing to make a difference. Instead, Balchowsky drilled the suspension parts so that if the car hit the wall, which it did, the suspension would bend and absorb the impact

Ol' Yaller II	
Wheelbase	94 in.
Length	na
Tread	56 in. front, 55 in. rear
Weight	1,940 lb. with 10 gal. of fuel
Engine	Buick V-8
Displacement	401 ci
Claimed power	305 bhp
Top speed	na
0-60 mph	na
Quarter-mile	na
60-0 mph	na

Ol' Yaller II, which was neater in action than Ol' Yaller I if not exactly neat, was driven by no less than Dan Gurney at Riverside, 1960. It had a cleaner body but retained the brushed-on yellow paint and the wide whitewalls. Road & Track

Ol' Yaller II was typically sturdy in the Balchowsky style, with ladder or truss frame and production suspension components adapted to the task. Road & Track

Something of a family portrait, as Ol' Yaller I lines up at Palm Springs, 1960, with a Birdcage Maserati in the center and Ol' Yaller II on stage right. Road & Track

and protect the frame, which was much more complicated and time-consuming and expensive to fix.

The body for Ol' Yaller II was aluminum, shaped over small steel stringers that conformed to one sort of general line for the car. Actually, it wasn't so much a body as it was a set of panels that covered the chassis. Each panel was attached to the stringers with Dzus fasteners, the kind that release with a quarter turn. People used to laugh, Balchowsky said, until they watched and learned how much more quickly the car could be tuned or serviced or even repaired with the pertinent panel removed.

The rear suspension was live-axle located by two semi-elliptic leaf springs. The rear shackles were linked in a sort of anti-roll bar that Balchowsky said was "self aligning ... it floats as much as a quarter of an inch."

No magazine officially tested Ol' Yaller II, although *Sports Car Illustrated* did a detailed analysis. Its writers were impressed with just about everything they found. The only components not at least average were the brakes. Balchowsky used finned aluminum Buick brakes because he believed in them and wasn't yet convinced of discs, and because until he got a really determined driver nobody knew that the brakes weren't as good as the rest of the car.

That driver was Dan Gurney. Right, *that* Dan Gurney. The local hero had graduated into Formula One but still found time to come home for visits and to drive the pro races for Balchowsky. This was the man who drove the big Ferrari sports cars that raised the level of involvement and expertise, and who could have driven any sports car he chose.

That Gurney picked the homely Ol' Yaller II, which carried a Famous Potatoes license plate from Idaho and was driven to the races and to the store by Balchowsky's wife, Ina, is all the comment needed.

Or maybe not. Balchowsky played the rube but was a cosmopolitan man, and knew much more theory of engineering and racing than he liked to let on. Nor did he object to the publicity. He'd learned early that someone who has a pithy or meaningful comment makes the papers more often than does the chap whose comments stop at "Hi, Mom."

Gurney ran a few fast laps at Riverside when the car was completed. He was close to the record and said he could break the record and the two-minute barrier, if only the brakes would hold up. "This is as good a car as I've ever driven," he said on that occasion, "it's as comfortable as a baby buggy."

In the Riverside Grand Prix a few weeks later Gurney had third-best time and led the race until the crankshaft broke. The brakes faded early but didn't cost Gurney much time. The crank broke because the harmonic balancer slipped, and with it the timing. Balchowsky adapted the balancer from a Ford six, and that problem never arose again.

Thing is, the success enjoyed by the original car never returned, either. Ol' Yaller II had one major victory, an SCCA national in which Bob Drake defeated "The Terror of the West Coast" (quoting *Road & Track*), a Birdcage Maserati.

More escalation had occurred. Reventlow had sold his sports car Scarabs in order to focus on a new project, a Formula One car. The buyers were well-to-do sportsmen who had the time and inclination to keep the cars in contention, which they did. Against them were arrayed a good assortment of well-financed and equipped Listers, including a new streamlined version, and a flock of Birdcage Maseratis.

These last were a new trend. Maserati had fielded some monsters, huge V-8s with more power than chassis, but they couldn't match the Ferraris and Listers. So the Maserati factory went the other way, with a delicate-looking little car that used hundreds of tiny tubes, intricately plotted and located so as to give maximum strength for the least weight. This tiny terror was powered by a two-liter inline four, with less power than the bigger cars had but with an equal ratio of power to weight and better handling and braking. And anyway, the new tracks didn't have much in the way of straights, so if you came off the corners strongly, and stopped better and cornered better, sheer speed didn't matter.

Between the Birdcage, the Scarab and the Lister there wasn't much that the homegrown crowd could do in the big leagues, not at the front of the pack anyway.

We'll hear from Balchowsky again, though.

Bocar XP-6

Bob Carnes had also been busy. The crew from *Sports Car Illustrated* went back east to Colorado late in 1959 to try out Carnes' latest model, the XP-6 (X was experimental, P was the hoped-for production and 6 was the sixth model Carnes had made). Carnes had energy and intelligence and faith. He'd sold a handful of early cars: five XP-4s and five XP-5s.

The *Sports Car Illustrated* test said, "Based on road and racing experience with all these cars, Bob Carnes has laid out the XP-6 with the express goal of challenging the current best in sports car racing—the Scarabs—on their own terms."

Carnes planned to do it with power. The XP-6 was mostly the XP-5, except the wheelbase had been extended 14 full inches and the space used for a supercharger, a 4-71 off a GMC diesel, driven off the front of a Chevrolet V-8.

Some odd notes appear in the text, or perhaps some unorthodoxies in the tuning of the engine, because the compression ratio of 9.5:1 is several points higher than most people would use or could get away with—with a supercharger packing compression on top of that. And the engine was a modest 283 ci, while other builders were doing the easy stuff like boring and stroking. Carnes rated the engine at 400 bhp, which could have been correct if the boost was high enough.

Bocar XP-6	
Wheelbase	104 in.
Length	na
Tread	54 in. front, 52 in. rear
Weight	2,290 lb. curb
Engine	Chevrolet V-8, supercharged
Displacement	283 ci
Claimed power	400 bhp
Top speed	167 mph
0–60 mph	6.0 sec.
Quarter-mile	14.6 sec. e.t.
60–0 mph	na

The XP-6 used a multi-tube frame, with three sizes of steel tubing. It looked like a heavy Birdcage and weighed 154 pounds, which was reasonable. Front suspension was a tubular front axle, chosen because it worked at Indy and road racing circuits were as smooth as Indy. A Chevrolet rear axle was located by lower trailing arms and a trailing upper A-arm. This car had left-side steering and the engine was offset to the right, for balance with driver only; the test writer said he could tell.

Carnes used his own fiberglass body for the XP-6 (and the earlier XP-5) but said he was reluctant to sell just the bodies because he hoped to produce complete cars, finished to a high standard, and didn't want the buyer of the complete car to be shamed by sloppy work from the kit buyer.

Time for some harsh words. The *Sports Car Illustrated* test was not a critical test. Carnes was an interesting man, a bit of a maverick when it came to dealing with the SCCA, and he was willing to tackle the big jobs. So *Sports Car Illustrated* writers put a good face on everything they saw.

Consider the listed top speed. The specs say 167 mph, assuming 7500 rpm. That works with the gearing, it's just that the rev limit for the test was 6000 rpm. Furthermore, the test was performed at Continental Divide Raceway near Denver, one mile up. Altitude reduces performance, said the magazine, and that's true. But it's also true that supercharged engines, which are artificially packed with air and fuel, are less affected by altitude than are engines filled by atmospheric pressure alone.

The test also said there was a problem with traction off the starting line and that hampered the acceleration figures. The thing is, other magazines testing other cars with more weight in front, off the driving wheels, and with less power claimed and with more weight got better figures than *Sports Car Illustrated* did for the XP-6. The test engine simply didn't produce 400 bhp, not by the figures shown.

Bocar XP-6 was mostly an XP-5 chassis and body except that 14 in. were added to the wheelbase and a supercharger was added to the engine, sitting (logically) in front of the engine and using the added length of the chassis. But the power didn't add as much as the extra length took away. This shot is from Pikes Peak, where the Bocar was third in class and old nemesis Ak Miller won again. Road & Track

In the event, the XP-6 did not defeat the Scarabs. (The Scarabs defeated themselves, as we'll see.) The XP-6 didn't win major races, nor did Carnes become a manufacturer. Like many an enthusiast before and after him, he learned the hard way that enthusiasm and technical knowledge aren't enough.

It was a gallant effort though, and for years afterward people read these stories in study hall and carried the figures with them as proof that the impossible could be done if you cared enough to do it.

Scarab Formula One

Sigmund Freud wrote that all wounds are self-inflicted. His reference wasn't to war or sniper fire but to the theory that our own thoughts and fears and perceived shortcomings cause us more pain and suffering than all the sticks, stones and names our enemies can hurl at us.

This is the story of a team that had almost everything and failed for the lack of something only they could have provided.

When the motorcar was invented, nobody knew or even cared where the powerplant should be sited. Some experimenters put the engine under the seat, some in front of the people, some aft. One clever chap built a motorcycle that consisted of an engine riding in a trailer behind a bicycle.

The car, racing or road, evolved with the engine under the hood in front of the cockpit and driving the rear wheels by means of a gearbox aft of the engine and a driveshaft down the middle. People sat behind the engine and next to the transmission and driveshaft. This worked fine, in all manner of cars, for fifty years.

There were experiments. Some front-drivers were produced from the beginning. Dr. Porsche was probably best known for his Auto Union Grand Prix cars, with giant engines in the back, and with swing-axles and a handful of eccentricities that made the cars more than a handful under power. The rear-engine Tatra was like a big pre-Volkswagen model. Dymaxion inventor Buckminster Fuller built a series of cars with the engine *and* the steered wheels in the back. More conventionally, Frank Kurtis built several rear-engine Indy cars before and after World War II.

Dr. Porsche was his own man and did what he believed whether it worked or not, but that didn't cause people to copy the Auto Union. Kurtis was pragmatic: If it worked he used it, and if it didn't work—which the back-motor cars didn't, not with the suspension and tires of the day—he put it aside and built what did work. Fuller was a visionary with small regard for what people would accept. He reckoned they *should* do what was right and if they did what they wanted instead, hard luck on them.

In short, front engine and rear drive was what everybody with any sense did, and everything else was out on the fringe.

Then came John Cooper. He had a small shop and built a few conventional sports cars after the war but became most involved with an idea for inexpensive racing: single-seaters with 500 cc engines. As the idea developed, the cars in the little man's class used motorcycle engines right behind the driver and driving the wheels with chain. The engine was upright and so was the occupant, and shorter chains are better than long ones... the engine went aft because that's where it fitted best. Cooper, along with peers and rivals, used suspension and other bits from Europe's ubiquitous Fiat 500; that's where the transverse leaf springs all came from.

Some of the small makers folded, some went into more conventional machines, but Cooper believed in rear engines and kept making them larger and larger, from Formula Three to Formula Two to Formula One. Two points here: First, back motors snuck up on

The Scarab Grand Prix car in early testing. Beautiful lines, wonderfully crafted... and woefully obsolete by the time it got to the track. Road & Track

107

Grand Prix racing in the mid- to late-fifties. Second, John Cooper didn't have money and expertise, so because he couldn't do it better, he did it differently. Keep that in mind.

Popular wisdom was dragged along behind the rear engine. All the books said it wouldn't work, then they said it wouldn't work with a big engine and then, when all the racers had motors in the back, they said everybody should have known it all along. Tricky stuff, this status quo.

Across the ocean and the continent, on the West Coast, Lance Reventlow and his handpicked, expertly backed team of craftsmen had conquered American sports car racing. They sold the sports cars, two of the three at least, to teams that went on to win on their own. The roadsters were sold because Reventlow was determined to hand the world the sort of lesson he'd just handed the SCCA and peers. He was going to build a world-beating Grand Prix car.

He didn't do it. But before we examine the car and the design and the campaign, we can draw a conclusion: Reventlow's crew failed because they knew they could do it better, and thus didn't need to do it differently.

The record shows that none of the major teams, such as Ferrari or Vanwall or Maserati, liked rear engines at first, and they were forced to change because the new little cars beat them. The same thing would happen at Indianapolis a few years hence. And to prove the point, in 1958, when Reventlow and company were laying down the basic form of their car, Aston Martin in England was doing the same thing—

The Scarab's detailed configuration was nicely thought out. That's Daigh at left, Reventlow behind the rear wheel. At lower left is the crankcase, with the air horns above because the engine was lay-down style; on the left is the crank; and on the right is the cylinder head. At lower center is the oil cooler with scoop, and just in front of the left rear wheel is the gearbox, sitting next to the driver so the car is balanced evenly front to rear. The driver sits lower than he could straddling the driveline. Road & Track

and was also going for the classic front-engine, rear-drive configuration.

Reventlow's company, Reventlow Automobiles Incorporated or RAI for short, had talent like Dick Troutman and Tom Barnes, Phil Remington, Chuck Daigh, Ken Miles and Warren Olson, all proven experts. They had Leo Goossen, who was the man behind the evergreen Offy engine, to do the powerplant. They drew up a standard Grand Prix car. At the time they reckoned to field an attempt in 1959, to see how it went, and then arrive in force for 1960, the last year of the 2.5 liter formula then in effect.

It didn't work out like that.

The engine wasn't much of a problem. Goossen knew what worked and so drew up an inline four-cylinder engine, double overhead cams, two valves per cylinder. It was roughly like the Offenhauser except it had provision for a clutch and gearbox and generator and had a cylinder head with valve guides and water passages laid out so as to keep in shape despite using gas for fuel, the shortcoming that kept the Offy from being a good road racing engine.

The frame was multiple steel tube, the body panels were aluminum, a better material for handworking and low production (as in one or two examples). The engine was tilted far to one side and the driveshaft was offset so the driver could sit next to it, rather than above or astradde.

There were problems from the beginning, but not many with the engine, which used two spark plugs per cylinder and had the latest in fuel injection. It had a stronger lower end than the Offy did, so it would hold up to stress even though it weighed more than hoped for. The comparison was with the 2.5 liter Climax engine, as seen in the backs of Coopers. That engine had roughly the same bore and stroke and the same valve size and while there were claims of 290 bhp from the Climax, the teams that used it said actual power peak was 240 bhp. After not a lot of development the Scarab engine was showing 230 bhp, so that seemed right on course.

The cylinder head for the Scarab four-cylinder engine had twin overhead camshafts, with a window for the cam drive at left. Combustion chambers were hemispherical, with deep half spheres that required a high domed piston to achieve a high enough compression ratio. This gave a poor shape for actual combustion, thus the twin spark plugs at the top center of the dome. There were only two valves per cylinder because the multiple-valve head was just then being re-invented by the Japanese motorcycle racers. This was a beautiful, expensive and inadequate engine. Road & Track

At the Grand Prix of Monaco, 1960, Lance Reventlow raises his hand to indicate he's coming into the pits. The team took their new cars overseas and learned that they were out of the running because the cars were out of date before they arrived. To be fair here, note that the Ferrari at lower left is also front engine, but Ferrari could afford to switch and Scarab couldn't. Road & Track

Tom Wolfe wrote that a pilot in trouble shouts, "I've tried A, I've tried B, I've tried C... What do I try now?" That looks like what Reventlow is asking here. Note that the tilted engine puts the crank and driveshaft on the left and the driver is offset to the right. Road & Track

Reventlow's goal was to build the all-American Grand Prix car. The first surprise came when few American suppliers showed any interest in the patriotic project. None of the firms making shock absorbers came forward with help on the suspension, for one thing. More serious—the team had the expertise to rework their own shocks—was the lack of domestic brakes. None of the American brake manufacturers were ready for, or much interested in, working with the team. The Scarab builders first tried an innovative drum brake with an expandable sort of bladder that pushed the segments of brake lining against the drum, but the brakes worked fine on the test bed and erratically on the car. So they took the easy way, and opted for proven Girling discs, from England.

Problems began with a lack of space. The smaller a Grand Prix car is, the better, so when the car was drawn, it was made as compact as the pieces packed inside would allow. Initially the driver sat next to the differential, with the gearbox just ahead of the differential, well back from the engine.

Suspension was independent, front and rear, so it came as a problem when it was discovered that when you tried to put the differential and the driver between the rear wheels, there wasn't enough room for the halfshafts to go up and down.

The rear disc brakes went outboard, where it was cooler but where they added to unsprung, uncontrolled weight. The gearbox went forward. That threw off weight distribution, so the engine had to be moved back, and that meant the frame needed to be reworked.

Then came handling. With fuel and driver the weight distribution was forty percent front, sixty percent rear. Previously racing cars of this configuration had been closer to fifty-fifty. The major shock here was that even with, or perhaps despite, the weight bias to the back, the front slid first. Some of this understeer came from the tires. The rears, which did the driving forward and had plenty of power to transmit, were naturally larger than the fronts, which means they stuck better, to the point that when the driver put his foot down, the front slid even more. Using larger tires on the front made the steering go all heavy and vague. The usual fix would be swapping or adjusting spring and anti-roll bar rates, which is what the team did. However, when the car was fine on the straights and fast bends but plowed off course on the slow turns, adjusting the chassis so it would be slick into the turns and could come hammering out with full power and the front end sticking made the car wander on the straights and plow on the fast stuff.

None of this was new, either to this team or to any team developing a new car. But because the suppliers weren't there and didn't come through on schedule, and because one of life's rules—Everything takes longer than you think—was in full force, the 1959 season was over before the car was halfway ready. Even then, the builders had to suspend testing simply because the time they had left before the 1960 season was needed to bolt things together.

Road & Track ran a cover picture of the car under test, implying that there were facts and figures inside. There weren't. All the magazine had time for, or maybe had the heart for, was a review of the tribulations and tests to date. The article ended as the boat steamed over the horizon: "So it is that after many delays the Scarab finally gets into action; about 200 lb. too heavy and at least 10 bhp short, but still a remarkable effort in light of the difficulties encountered. Not the best, but still very promising."

Two months later came the race report from Monaco. BRM and Ferrari arrived with new rear-engine cars, Cooper and Lotus had already gone that route, several entrants were using Cooper chassis with Maserati and Ferrari engines. In the back, of course.

They had held the revolution without the Scarabs.

Stirling Moss put his Lotus-Climax on the pole, with a time of 1:36.3. The qualifying cut was 1:39.1. Chuck Daigh worked down to 1:47; Moss himself tried the Scarab and turned 1:45, and the clock ran out with the Scarabs unqualified. They were too slow at the Dutch Grand Prix as well. At Spa, the Belgian Grand Prix, there was a small entry but the Scarabs were still off the pace. Jack Brabham did 3:50 in the new Cooper; Phil Hill turned 3:54 in the outdated (front-engine) Ferrari; and Chuck Daigh, who by now had a reputation as a driver of Grand Prix caliber, was clocked at 4:09. If it was any consolation—and it surely wasn't—the Aston Martin team didn't bother to attend. Several engines blew up during practice at Reims, for the French Grand Prix, and the team packed up and went home.

They were going to switch, Reventlow said, and would return with a lighter, rear-engine car, while they'd left engine parts in England for wizard Harry Weslake to examine and improve.

By chance the 2.5 liter formula was ending in 1960, as the FIA had decreed a new limit of 1.5 liters. The FIA later came through with a new class, Intercontinental, which had a 2.5 liter limit, by no chance the very size of all those expensive and outlawed racing engines. There was talk of persuading USAC to have racing for this class, on road courses and even at Indianapolis, so the United States would once again be a part of the rest of the racing world. There was also talk of shipping a little Cooper over to Indy, for testing.

Back at Reventlow's shop, meanwhile, some experimentation was being done on switching for the new formula with the old car, or going another giant leap with a small-rear-engine Grand Prix car. After debate, the team went forward one step, with a rear engine, and back one step, with a sports car design.

Years later, Phil Remington said that the supposedly overbuilt engine block flexed when more power was extracted from it, a self-defeating flaw. Considering the project overall, he said, "It was a waste of time to go. We were hopelessly outclassed."

The state of the art

Early in the season, back in the United States at least, 1960 was called The Year of the Birdcage, meaning the year of the deft little Maserati, formally known as the Tipo 61. This was an evolved model by that time, with a long aerodynamic tail and high windshield that began well forward of the normal cockpit.

The race of the year was at Road America. The pioneer racing around the town of Elkhart Lake, Wisconsin, had been replaced, thanks to faith, work and capital, with racing on a lovely, scenic, challenging road circuit. USAC-FIA put on the first professional race meeting there, and attracted top teams from east and west and points between. There were two rows of Maseratis, one driven by an engineer-oilman named Jim Hall, and there was a salesman named Roger Penske with a Porsche RSK. Carroll Shelby, who was supposed to drive Formula One for Aston Martin, spent the year in the United States and on this occasion was aboard Ol' Yaller II, with the largest engine in road racing. There were two refurbished Scarab sports racers, bored from 339 ci to 348 ci, and a fleet of Ferraris, plus, as *Sports Car Illustrated* said, "enough additional Corvette powered specials to fill the GM proving grounds with glee."

Other drivers included three from the Indy pro ranks and a couple of true amateurs driving their own creations, Chevrolet-powered homebuilts named Echidna; *Sports Car Illustrated* taught us to pronounce the odd name by rhyming it with Heck did ya?

The first 100 miles of the feature were nip and tuck, with Ol' Yaller in front of Scarab until both broke. Then it was Jim Hall against Billy Krause in a

Ak Miller, the very model of an enthusiast, on his way to winning class at Pikes Peak in his Devin-Chevrolet. Road & Track

Billy Krause, a tough kid who worked his way into the top ranks, hammers on his Chevrolet-powered D-Type Jaguar while Bob Drake in a Birdcage Maserati waits for his chance... which he'll get when the Jaguar breaks. Road & Track

This tiny version of a Lister is Martin Tanner's T-3, powered by Saab. Tanner, a gifted craftsman, made the aluminum body himself, and adapted the front-drive Saab engine to drive the rear wheels. Road & Track

Chevrolet-powered D-Type Jaguar, with Jim Jeffords driving a Maserati in second until he got the lead, which he kept while . . .

The Maseratis finished one-two-three because the bigger cars broke. In the long run the top finishers won a couple thousand, the others got a couple hundred, and the fans and promoters went home happy.

This was the first of a series of professional races. The Meister Brau Scarab proved that sponsorship was worth having and worth bestowing, which had been the tough one until now. Road racing was becoming a marketing tool as well as a way to earn a living.

The consequences were not yet fully understood, but the 1960 Road America was a watershed.

Martin T-3

There was still a place for the home craftsman, for example Martin Tanner, whose class H Modified was described earlier. He'd switched from Crosley power to Saab, an odd little three-cylinder 750 cc two-stroke built by a Swedish aircraft company best known for doing things the way nobody else would have dared to do.

The T-3 was a clearly evolved version of the T-1 and the later T-2, which was sold because Tanner didn't like its looks. The T-3 had no such problem. The body was aluminum, hand-shaped by Tanner, and looked like a cross between a Lotus sports and a late-model Lister: fiberglass had its place but aluminum, done right, still worked.

Tanner's frame was the by-now-standard multi-tube, with high sides so the upper and lower tubes were far apart. The tubes were aluminum as well, something Tanner could get away with in a car this small and light. The frame was stiffened by a magnesium floorpan and alloy firewall, riveted to the frame tubes. Front suspension was based on the classic Fiat 500, with shock-springs from a motorcycle sidecar.

The live rear axle was located with radius arms, leading rather than trailing so there'd be more room for the driver.

The major engineering problem on the car came because the Saab was front drive and the T-3 was rear drive. Tanner kept the Saab clutch but surrounded it with a bellhousing made from aluminum sewer pipe, no kidding. He fabricated the housing and welded it to a Fiat 500 transmission.

The engine was straightforward, sort of. Saab was rallying its cars, so it had tuning books and kits for the engine. Tanner raised the compression ratio in the combustion chamber and again in the crankcase. Two-strokes breathe the fresh fuel-oil-air mix into the crankcase from the carburetor, then compress it so when the piston is in the right place and the ports are open, *whoosh*—the charge goes in one side of the cylinder while the exhaust is expelled out the other. The engine used three carburetors, one per cylinder, and cranked out 55 bhp.

No magazine did a formal test of the car so all the figures available are simply what the builder said they were. Tanner weighed the T-3 at an incredible 699 pounds dry. The class rival then was the OSCA, with a shipping weight of 948 pounds. Throw in, say, 60 pounds of fuel and 150 pounds of driver and the water, battery acid and so forth, and you'll get 1,000 pounds for the T-3, 1,250 for the OSCA. Claimed power for the OSCA's twin-cam engine was 70 bhp, so the special enjoyed a power-to-weight advantage over the factory machine.

Tanner could and did beat the best OSCAs straight up. There was a place for the home craftsman, albeit Tanner was learned in theory as well as experienced in the shop.

Echidna

At this point in time, the main events were being contested by some remarkably sophisticated cus-

Echidna

Wheelbase	93 in.
Length	na
Tread	52 in. front, 50 in. rear
Weight	1,928 lb. dry
Engine	Chevrolet V-8
Displacement	339 ci
Claimed power	340 bhp
Top speed	145 mph est.
0-60 mph	4.5 sec.
Quarter-mile	na
60-0 mph	na

tomer racers—Listers and Maseratis and Porsches—and by no-holds-barred specials like the Scarabs and the later Ol' Yallers, which were much better planned and executed than their body panels or press agents let on.

Then there were the specials from Minnesota, the Echidnas.

Sports Car Illustrated summed up this team's efforts, and secrets, perfectly. The title of the article was "Three Who Thought Things Through."

The three were John Staver, a partner in a foundry; Ed Grierson, a sales engineer; and Bill Larson, an optometrist. They lived in northern Minnesota where, as the magazine said, the winters are long and deep and there's plenty of time to work in the shop, assuming a good shop heater.

Staver was the sponsor, Grierson the builder and Larson the pragmatist. All three had raced, stocks in Larson's case and production Jaguars and Corvettes for all three. In 1957 they came to a conclusion common in those days: If you can go fast in a 3,000 pound Corvette, how much faster could you go in a 2,000 pound car with all the Corvette good stuff?

Unlike most who wondered, this trio had the means and skill to find out. They began with the frame from a 1956 Chevrolet sedan because such frames were common and inexpensive and much better engineered than people realized. The frame was shortened and narrowed and the various cross-members and gussets were relocated. It was, they reasoned, quicker and more practical than forming all the tubes and sticking brackets on them and working out the location devices for the engine and so on, which could now be bolted on, and anyway, the finished frame was only a few pounds heavier than a ladder or multi-tube frame would have been.

The same thing went for the front suspension, which was mostly Chevrolet. They used Chevy steering at first, although later in the project they switched to Morris Minor rack and pinion. The team was happy doing things their own way, such as running the steer-

This Echidna was one of three built by a talented trio of amateurs from Minnesota. The builders weren't fussy, as you can see by the taped body damage on the right rear and the use of a knock-on rear wheel and bolt-on front. The body was Devin and the frame was cut-down and reworked 1956 Chevrolet sedan. The car sat high because the builders believed there was less wind resistance from a high car than there was if you had to shove a low car through air compressed between the car and the ground. Road & Track

ing linkage through a hole cut in the frame because that's where proper geometry said it should be. Or deciding that because most road racing turns were to the right, the left front wheel could be decambered by several visible degrees.

The Chevrolet rear axle was narrowed, on one side only, because engine and gearbox were offset to the right to balance the weight of the driver on the left: note once again how often clever thinkers do different things to solve the same problem. The axle was suspended by leaf springs, just as on the sedan, but with trailing radius arms for fore-and-aft location and by a Watt link for side-to-side control.

The Watt link is a geometric marvel. To fully understand how it works, know here that it comes from James Watt, the man who invented the Steam Age. Visualize two parallel rods, one high and one low, one left and one right. Their inner ends are directly aligned vertically; that is, the right end of the high rod on the left is directly above the left end of the low rod on the right. Place a pivot point in this vertical line, halfway between high and low. Attach a short, vertical rod to the inner ends of the two horizontal rods and pivot the short rod on the central pivot while letting the long rods swivel at their outer ends.

The linkage lets the central pivot, or whatever is attached to the pivot (for instance, the rear axle), move freely up and down, and tilt side to side, or lets the outer location point of the horizontal rods go up and down, or tip, while *the central point is always central.* The central point can't move side to side. So, if you have the outer points on the frame and the central point on the differential, the axle is perfectly controlled laterally, and perfectly free in the other directions.

Who did this first? The name has been lost. We'll see this linkage again, but its use when and where most builders had the classic Panhard rod shows that the Echidna team came up with a better way. (With the Panhard rod, one horizontal rod runs from the frame on one side to the axle on the other, resulting in lateral motion along with up-and-down motion and in a car that handles differently right versus left. The Elva Courier had just such a rear axle linkage.)

Brakes began as Corvette, the drums that used cerametallic linings against finned drums. This was semi-standard for racing cars of the day that hadn't gone with disc brakes.

The trio didn't believe in discs. Instead, Staver being part owner of a foundry, they kept the Corvette backing plates and linings but used their own highly finned drums. The argument against cerametallic linings then was that they didn't work well until they were warmed up, and that they wore out when they got hot. The Echidna team said they'd learned how to keep the brakes at optimum temperature, and that they'd found the cerametallics working best when they looked wrecked. The Echidnas did stop at least as well as their rivals—another case of doing a thing their way and having it work.

The roll bar, by now required for racing despite protests from those who believed they could tuck under the dashboard when they felt an inversion coming on, was also engineered to provide structural stiffness, to be part of the frame.

The Echidna's bodies, on the other hand, were Devin shells. They covered the machinery and did nothing else—by intent. The designers didn't want to try controlling that which was beyond their control.

In the same manner the Echidnas always had a riding height visibly above what most sports racing specials used. Conventionally, the lower the better. But in this case the theory was that specials (and factory cars as well) got unstable at top speed in part because the air was trapped and turbulent beneath the car. The car rushing over the ground at minimum clearance was floating on its own compressed air. So, the easy way out was to leave the car high and not compress as much air. Again, it seemed to work.

There were three Echidnas, one for each partner. Staver's car was first and raced in the SCCA's C Modified class its first year. When the other two cars were completed they got 283 ci engines, which Staver began with, and Staver's car grew to Scarab size with a 339 ci Chevy V-8.

At the time, the SCCA classes went A, B and C Modified. Class A was virtually unlimited, the former Unlimited class having been dropped when four-wheel-drive sprint cars and old classics were no longer needed to pad the program. In 1960, just after the Echidnas did their best, A and B were dropped and C Modified, three liters and up, was the biggest class. (Why the SCCA dropped useful letters like A and B and retained the others, down to H, was never explained.)

The Echidna engines began as stock Corvette powerplants and didn't change much. They used stock heads and pistons and the off-the-shelf Duntov cam. They used carefully and expertly modified Rochester fuel injections, despite the sure knowledge in racing circles that the central fuel injection, as seen in production racing Corvettes, wasn't any good. It worked for this team. Their secret—if secret is the right word and it's not, because they told everybody how they did it—was careful assembly and matching of all the parts and keeping the clearances optimum, as opposed to maximum.

The Echidna boasted some impressive speeds— with a note in that the *Sports Car Illustrated* staff apparently tested the class B car, and there's a curve showing acceleration, but the staff didn't actually drive it. The 0-60 time is taken from the chart; no figures were given for that time to speed, nor for the quarter-mile. Assuming the claimed power and weight to be accurate, and relying on the team's track record as well, we can guess that the B-engine Echidnas would have run neck and neck with the Scarab on

A Buick-powered special with body by Hagemann (no it's not an Ol' Yaller; sure looks like one, though) leads a Porsche RSK. The big car with big engine and little car with little engine were often this closely matched. Road & Track

Walt Hansgen, national champion in C Modified, and his Cunningham-stabled Lister-Chevrolet. This is the later version of the bad machine seen earlier, and it was much more aerodynamic if not as threatening to see in one's mirrors. Road & Track

a normal day for them both. Beyond that, the magazine quoted a good outside driver, a man who co-drove for an endurance race, as saying that the lines of the car were comfortable and forgiving and predictable despite some of the notions like the decambered left front wheel and the mild engine and much-maligned brakes.

The odd part here is that the best comparison, the results of the races, isn't always useful.

From 1959 through 1960, there were the SCCA and USAC nationwide. There were big pro races and national amateur races and sprints-versus-endurance events. There were Cal Club on the West Coast and Nassau Speed Week as an outside event off the East Coast. SCCA national titles were awarded to the drivers who scored the most points in a long season, in races across the United States.

So, it happened that Walt Hansgen, who drove a Lister-Jag and then a Lister-Chevy for the Cunningham stable, was national champion for the big modifieds owing to his skill and his equipment and his team's willingness and ability to crisscross the country to collect said title.

Reventlow did the same thing, and later, when the sports cars were sold, so did the Meister Brau team when they owned the Scarabs. In the Reventlow days, the team made sixteen pro or national starts and won twelve of them; under the brewery's banner, they made sixteen different but equally important starts and won twelve times then, too.

Meanwhile the Echidnas were driven and sponsored and maintained by guys who had businesses to run and who had no sponsors willing to believe that it was worth spending $100,000 for a $5,000 purse and your name in the paper. These men stuck close to their remote homes, and in *their* thirty-five starts they got eight class or outright wins and were in the top three eighteen times, or nearly every other time they fired the engines in anger.

Not a bad balance. By this time the SCCA let you drive professional events if you didn't take the money and the USAC let you not take the money; no fools

Briggs Cunningham—sportsman, driver and team manager—watching his team in 1962. Road & Track

they. During all this the big amateur team with professional cars, the sponsored team with no-holds-barred specials and the team of three guys who thought things through were each capable of sending the other two home on their shields.

As racing reporter Pete Lyons wrote, good racing is when you don't know who's going to win.

Chapter 6

At the Peak 1965–1966

The specials and the sport get so much better that one
has no place for the other

Start with a finish, a terminal finish.

When last seen Ol' Yaller the First was parked in Max Balchowsky's yard, with engine by Balchowsky and the rest of the car owned by Eric Hauser, a broker in real life and a good driver for the joy of it. When the partners went their amicable ways, Balchowsky naturally got the engine and Hauser the car. Hauser did the normal thing and got a modified Chevy V-8 for the car and campaigned it in that form.

The old special, which began life in 1953 remember, with power (sort of) from a side-valve Plymouth six, was still in the hunt early in 1960, as it led the big race at Palm Springs, California, only to retire with gearbox trouble. The car's last win was late in 1960, when Hauser took the amateur race preceding the pro race at Laguna Seca.

Then came Cal Club's Riverside race, early in 1961. Hauser arrived late and didn't get much practice, so he finagled extra time. The engine seemed to be down on power, and would propel the old warrior only 120 mph down the straight. Hauser decided that if the car was that slow he could get it through the first turn, an over-the-crest 100 mph left, flat out.

He was wrong. Ol' Yaller slid off the outside of the turn and into the guardrail, careened off the rail across the track and up the other side, turning end over end until it came to rest upside down with Hauser, his worst injury a cut lip, hanging from his harness. To look at the tangled mess was to know the car would never race again. In a fateful sort of way, it seemed proper that the old beast's incredible career ended as dramatically as it did.

Elsewhere, things were moving right along. At the Laguna Seca meet that saw Eric Hauser score Ol' Yaller's last win, Stirling Moss took the professional event at the wheel of a Lotus 19, a sports racer with Coventry Climax power, a twin-cam four, in the back. Billy Krause won the Times Grand Prix at Riverside with a Birdcage Maserati, which replaced his Chevy-powered Jaguar. A crowd of 60,000 paid to watch Moss and even more went to Riverside, thanks to the *Times*' ceaseless promotion of its own event.

The SCCA had taken a sensible, if belated, step in re-classifying production cars. For a generation they were classed on the basis of engine size. The power one got from said size depended on price and the expensive car usually beat the less-expensive car; say you could afford a two-liter Triumph and I could afford a two-liter Ace-Bristol that cost twice as much, I won.

The new rules began with those who knew sitting down and figuring out how fast you could get, say, an Austin-Healey to go if you prepared it to the limit of the rules. What about the Porsche Super? the MGA? Then the classification committee assigned the cars in groups: A, B, C, D, production. When a particular model was faster or slower than expected, it was re-assigned. It was a good idea, carried out with care and fair play, and it worked. It also meant that racers on a budget could suddenly stand a better chance of winning with a production car, which meant they'd have less reason to build their own.

Parallel to that, the growing popularity of road racing, professional or not, created a demand for racing cars. The big ones came from Italy (Ferrari and Maserati), the middle was occupied by Porsche, and a host of small English firms (Lotus, Lola, Cooper, Elva and so on) provided the small racers with Climax engines.

They were good racing cars. The Riverside and Laguna Seca events attracted a flock of specials, as the magazines noted at the time. The thing was, they didn't have the best drivers and they didn't win the big races. The competition came from two sides: the big engines (specials for the most part) and the rear engines (Lotus 19 or Cooper).

There was potential for a middle ground. The little Chevrolet engine (little in the terms of 1949's V-8s) was being challenged. Ford had a V-8 that was smaller on the outside as well as the inside, for its smaller passenger cars and in due course the Cobra and Mustang. GM's middle divisions were sharing what they hoped was the engine of the future, an aluminum V-8 for the smaller Buicks, Oldsmobiles and Pontiacs. *Road & Track*'s gossip column said that Tom Carstens (of the Allard-Cadillac and HWM-Chevrolet) was busy on a Lotus 19 with a B-O-P (Buick-Oldsmobile-Pontiac) V-8. They expected 240 bhp from 3.5 liters, which would put the special right into the old power-to-weight contest with the Climax-engined cars.

Which makes the next step all the more surprising.

Comstock-Sadler

Because it was fair to criticise Reventlow and company for lack of vision, it's fitting to say that Bill

The Lotus 19 was a popular and competitive racing car. This one, driven by Jerry Grant, was built with a Coventry Climax four but later got a Buick V-8 and here had been stretched and fitted with a Chevy. Road & Track

The death of the original Ol' Yaller came at Riverside early in 1961. The car was demolished, driver Eric Hauser was unscathed but an era had ended. Road & Track

An offshoot experiment: Bill Sadler built this front-engine Chevy-powered single-seater just barely in time to do poorly at the US Grand Prix at Watkins Glen in 1962. The car was mostly a solo version of Sadler's Mark III sports racer. Watkins Glen Racing Museum Archives

Comstock-Sadler Mark V	
Wheelbase	90 in.
Length	na
Tread	50 in. front, 48 in. rear
Weight	1,475 lb. dry
Engine	Chevrolet V-8
Displacement	364 ci
Claimed power	370 bhp
Top speed	na
0–60 mph	na
Quarter-mile	na
60–0 mph	na

Sadler had vision . . . and the courage of his beliefs. He'd done the conventional specials, the TR-powered rear-drive and front-engine cars that grew new frames and engines and bodies. He'd even made a single-seater that began with the engine in front and then had it in the back.

Sadler was a man who knew which way the tide was running. He also had support from the family business and from other Canadian racing figures, notably national champion (in an Austin-Healey) Grant Clark.

Late in 1960 Sadler began building two new cars, one for him and the other for Clark. There was supposed to have been a second angel, but that deal fell through and the cost of the second car was picked up by Canadian Comstock, a Canadian construction firm that paid for the building and the racing in exchange for the model's proper name becoming Comstock-Sadler.

The frame was multi-tube, small diameter and mild steel. It was basically a space frame, with stressed panels, and according to *Road & Track* it weighed only 65 pounds while testing out at 3,000 lb-ft per degree (meaning that to warp it one degree out of true, you needed to apply 3,000 lb-ft of twist). Stiff, in other words.

The Mark V was light right from the drawing board and could save time and work by using Healey suspension in front, along with rack-and-pinion steering based on Morris Minor but housed in Sadler's own case. The brakes were Girling disc, MGA at the rear, Austin-Healey in front, and the coil-over spring-

The Comstock-Sadler Mark V, here under early test, was a newer and radical (for its day) sports racer with a Chevrolet V-8 in back and with only two forward speeds. Road & Track

The Sadler Mark V was clean, with narrow FIA-style cockpit. What looked like several steps in the right direction became, presumably for financial and mechanical reasons, the last Sadler. Road & Track

shocks were Girling, as seen on the Lister. (Sounds as if the old junkyard has been replaced by the import agency's parts bins, but just as in the old days, the components were good ones and it would have been foolish to make or buy special-order stuff when you could get equal quality over the counter.)

Sadler provided a wide car and narrow cockpit, with the driver on the right and the fuel tank slung pontoon style between the left front and rear wheels, outboard of the cockpit. That would have balanced the weight side to side, but only when the tank was half full, or full or empty: not a fatal flaw, but something one wonders about.

The rear brakes were inboard and the suspension was independent, with the halfshafts (driveshafts) serving also as the upper link and with radius rods from the hubs to the center of the frame serving as the lower links. Sadler had paid attention to what others were doing; indeed he'd stolen a march on Chevrolet with his rear-engine single-seater, and the Mark V's suspension was equal to anything being done at the time.

Ditto for the engines. One car got a modified-but-not-all-that-much 283 ci bored and stroked to 327 ci. The second car carried Chevrolet with a whopping 360 ci or 364 ci—depending on which account one believes. Compression ratio was 13:1 or 14:1, again depending on the account, and the monster had Cadillac connecting rods, a machined-from-billet crankshaft and huge valves, with induction through six carburetors.

The powerful engine was set so far back in the wheelbase, barely ahead of the rear wheels, that there was no room for a gearbox. Anyway, with all that power the car didn't need many forward speeds.

Sadler's single-seater had used an engine tuned to perform well across a wide rev band and had a clutch, a differential and just the one speed. The Mark V had two forward speeds. Sadler did it by using a Halibrand quick-change differential, as seen at the oval tracks and dry lakes. A Halibrand quick-change is an addition to a differential. Where the standard unit has the pinion gear turning the ring gear, the Halibrand has drive to a pair of gears and then to the pinion, with the gears aft of the ring. It's a quick change because the back cover comes off and you can juggle the paired gears and thus change the final drive ratio quicker than you can swap differentials.

Sadler extended the housing for the pair of gears and tucked in another pair. He used Halibrand gears to make them fit, and Ford synchronizers. The shift pattern was low-neutral-high and because all four gearwheels could be swapped, he could tailor his two speeds for the track—for slow corners and long straights and so forth.

The alloy body was three basic bits: front, rear and pan, with flop-down doors. It had provision for lights, as rules required, and looked tidy and modern. Nobody ever said anything about aerodynamics, which is a surprise because car speeds were rising into problem numbers about then.

Except that the Mark V didn't last that long.

The cars were put together at speed, tested to see if things fell off, which they didn't, and Sadler and Clark entered the Player's 200 at Mosport, Canada. This was *the* Canadian sports car race, with drivers like Stirling Moss versus the locals.

After the race Sadler said Clark's car, with the 327, got bent in practice because Clark was so busy

Get used to this man and machine, Jim Hall, acting as owner and manager, gives the OK sign to Hap Sharp as Sharp and the Chaparral 1 win the USRRC at Elkhart Lake in 1962. Road & Track

121

out-dragging Moss down the straight that he flew off the track. In the race itself, both cars had... gearbox problems. The unproven parts, as often happens.

There's a puzzle here. The race report says both cars handled "atrociously," while the magazine accounts of the cars only hint at shortcomings in that area, stating that the cars will be sorted out.

The next step, one month after the Mark V's debut, was for Sadler to announce the closing of his company. He went back to school, and never built another car.

Puzzling. Probably Sadler had raced enough to satisfy that particular urge. It can be outgrown, after all, and Sadler was always thinking of the next step. It's ironic, though, that the first person to effectively use the Chevrolet engine in the proper place, and to experiment with gearboxes or the lack of same, was to get so little in return for his daring. Especially when a man who was similar but different became a household name by exploring in the other direction.

Chaparral 1

Jim Hall also had a family business, but he came from Texas and the business was oil.

Hall's older brother Richard was a sports car fan and sometime driver. When Jim Hall got out of college he was sent to Dallas to look after the sports car dealership his brother had established there.

The agency was more like a sports car club and, while the bottom line was more often red than black, there were some interesting people in the shop. Hall got to know a good amateur named Hap Sharp, and arranged to supply Sharp with a Cooper.

Hall began racing. It turned out he was good. Really good. He began with an Austin-Healey, presumably off the lot. He was so much a novice that he didn't know he was supposed to be twenty-one, or so he said later.

Hall moved into larger and faster cars, Maseratis and Lister-Chevrolets, and did well. He decided to try Formula One, but learned that despite his record and

The first Chaparral was tidy and compact, with a space frame and suspension borrowed from the pure racing cars. Because Troutman, Barnes and Hall knew the configuration was obsolete they took great care to make everything light and correct. And the approach worked. Road & Track

connections, when it came to Grand Prix racing, "They'd only sell us last year's cars."

So he bought a Lotus 18 (a Formula Two car with 1.5 liter Climax) and he bought a Cooper Monaco (a sports car with 2.5 liter Climax) and he did what you or I would have done and *presto*: a certified Grand Prix car.

The United States Grand Prix round was held at Riverside in 1960 and Hall was there, with a Lotus Grand Prix car the Lotus team didn't remember selling or even building. He worked his way as high as third, was fifth on the last lap and finished seventh because the differential disintegrated. Hall could drive. He kept the car.

Hall's best Formula One finish came in 1962: fourth in the Grand Prix of Mexico. Even so, it was obvious that he'd never be allowed to buy a top car and he'd need several years to earn one.

Hall was determined. He went back to school, California State Polytechnic University, for a degree in

Five Chaparrals were built and three became customer cars. This customer car is driven by Harry Heuer of the Meister Brau team. Road & Track

Cast rear uprights kept suspension linkage and hub inboard of the wheel. The wheel was located by a reversed A-arm on bottom, lateral and trailing arms on top, with canted coil-over shock and anti-roll bar. The cover for the quick-change differential is visible at left, below the spare tire. Road & Track

Cobra: The production hot rod

Carroll Shelby was a world class driver, which did him some good. He was also a man who could persuade other people that what he wanted, they also wanted. That did us all some good.

Shelby retired from driving with a heart murmur, and the ambition to do something in and with the sport of racing. By the sort of happy chance that some make for themselves, he heard that AC, one of the smaller English brands, was about to go out of the sports car business. They had an archaic chassis: flexible rails with independent front and rear ends via transverse spring, sort of like an old Ford or Fiat. They had a lovely roadster body, inspired by the Touring-bodied Ferrari roadsters of the late 1940s. And they used to have an engine, a Hemi-head inline six. The maker of that engine, Bristol, wanted out of the car business.

Shelby knew all about the domestic V-8. He made an approach to GM, as in Chevy, and was turned down flat. They didn't need him, nor what he had in mind. So Shelby tried Ford, where by still another happy accident they had a sharp little V-8 just going into production, and they had a big sort-of sports car on the way, but they didn't have an answer to the Corvette.

Shelby said in effect, if you guys here could supply the ham, and then you other guys over there came through with eggs, we could have ham and eggs.

Gosh yes, said both sides, so Shelby, who had no capital and no manufacturing experience and no design or management training, went into the ham and egg business.

Along with his gift of gab, Shelby had an eye for talent. He signed up chaps like Ken Miles and they discovered that yes, the small-block Ford V-8 went right into the AC chassis. They did some beefing here and trimming there and additions elsewhere. AC and Ford approved the prototype, and the inspired name Cobra, and Shelby was in business.

This is a grand saga, a hero on every page. The story has been told in magazines and books and it never fails to inspire and instruct. Shelby is a wonderful character, crafty and cheerful and honest about both.

But the Cobra story isn't in this book. It's not in the book because the Cobra was a production car, produced in the hundreds, with backing from Ford. It was sold in dealerships coast to coast, and in other countries. True, on at least one occasion, Miles modified a stock-class Cobra just enough to make it a sports racer and then defeated the real sports racers in an SCCA national. Equally true, the Cobra was crude, with flexible frame and questionable suspension and no surplus fat.

But the Cobra wasn't made at home for fun. The Cobra tore up the track, but it didn't break new ground. It yielded to no man the claim of biggest engine in smallest chassis; but while the Cobra was special, it wasn't a special, any more than Shelby's team was anything except the Ford factory's best.

A lashing of Cobras through the esses at Riverside. The Cobra roadster was the best example of its type—the lightweight, high power, race-designed sports car—ever. Lovely car, brutal at the same time and it owned the top production class. But it wasn't built for fun, or at home, so it wasn't a special by our terms. Road & Track

Harry Heuer again, far left, at the SCCA national at Daytona Beach, 1962. Look carefully below the car's nose and you'll see a metal panel below the radiator intake. That's an early air dam, put there because the speed of the light car was beginning to lift the front and Hall and GM were beginning to search for a cure. Road & Track

engineering. He did a lot of thinking. His exposure to the rest of the world had taught him that you couldn't drive much better than your competition could push you, and in the United States there wasn't enough push. Hall: "What we needed was better cars. There was no question they were better drivers."

Back at that 1960 US Grand Prix Hall had chatted with Troutman and Barnes. They'd left Reventlow and were on their own. They and Hall agreed with Sadler that rear engine was the way to go, but that there was no transaxle strong enough to stand up to the Chevy's power.

Hall was willing to compromise and did so in a direction he and Troutman and Barnes understood. He agreed to back the builders in a project, a front-engine Chevy-powered special. Hall would buy the first two cars, and Troutman and Barnes would have the right to produce more and sell them, as demand warranted.

The car was named Chaparral, for the bird that is better known as the Road Runner. Great minds do move in similar directions: Hall and Sadler, working on the same problem at the same time, came up with matching ideas. The Chaparral was small, with an 88

Caught with their panels down, the Chaparral team was forced to drastically tack higher bodywork to the car so it could meet FIA requirements for the 12-hour Sebring race in 1963. Road & Track

Chaparral 1	
Wheelbase	88 in.
Length	na
Tread	50 in. front, 50 in. rear
Weight	1,479 lb., no fuel
Engine	Chevrolet V-8
Displacement	318 ci
Claimed power	325 bhp
Top speed	na
0-60 mph	na
Quarter-mile	na
60-0 mph	na

in. wheelbase. It used large wheels, 15 in. in diameter, at a time when power could easily overcome traction but just before Mickey Thompson proved what others suspected: that if you had the right tire compound, the more contact patch you put on the ground, the better the tire stuck. The wide wheel was a few years away in 1961, but the long footprint provided by the tall wheel was taking over.

Hall's engine was mild, a 318 ci Chevy with only three carbs. Target weight was 1,500 pounds, and because the car had a four-speed Corvette gearbox light and little meant it could be tuned for as much power as it could use, but no more. Low stress equaled reliability, or so the book said and eventually proved.

Front suspension was conventionally upper and lower A-arms, with coil-over shock-springs, rack-and-pinion steering and of course Girling disc brakes: the last by this time were something the homebuilder specified without question.

The fuel tanks were amidships, outboard of the cockpit and part of the aluminum body. Troutman and Barnes had no superior in this field and the body panels were perfectly formed over carefully detailed bucks. The car was very low and the highway patrol in any state would have come to a sudden stop when faced with the height of the headlights.

Rear suspension was virtually Formula One. The differential, again Halibrand with halfshafts, carried the brakes inboard and had huge rear hubs so the upper and lower links, the A-arms, could be given lots of travel and optimum geometry.

The result was a beautiful little car.

The first time out, Hall finished second at Laguna Seca. He and Sharp began campaigning, and did very well.

None of the magazines managed to do a full, timed test of this car. *Car and Driver*, which *Sports Car Illustrated* had renamed itself by then, did a cutaway and detailed analysis, but mostly its writers saw the car as an evolved, improved Scarab. They were probably correct, as Troutman and Barnes had done most of the work on the earlier car. They'd learned that the Scarab was too big and heavy; they'd learned that because the cars didn't break enough. Remember fifteen years before this, the first American specials could compete with the imported sports cars in part because they didn't have to be beefy. Same thing here.

Next, a point *Car and Driver* took for granted was the ability to tune. This may have come from Frank Kurtis. His (and rival) Indy and sprint cars came ready to be adjusted for the track and the conditions of the day. This began as dirt track technology. Dirt tracks change during the race, not just for the day, and the racers there are prepared for it.

Ted Peterson, who raced hot rods through the Canadian-American Challenge Cup (Can-Am) and was the owner-driver of Ol' Yaller IX in 1989, said that one of Balchowsky's secret advantages was that his cars, like the Kurtises before them, came with settings

Sharp in the Chaparral leads Heuer in the Scarab at Elkhart Lake, 1962. There does seem to be a family resemblance; Troutman and Barnes had major parts in each car, and what they learned with the Scarab they applied to the Chaparral. Compare the rear profile of the Chaparral here with the one at Sebring in 1963, when more aerodynamic work had been done. Road & Track

for camber and caster and toe-in and torsion bar positions and anti-roll bar settings. You could dial or click in more or less oversteer or understeer or alter the ride height to allow for bumps or the lack thereof . . . all this assuming the tuner knew what he was about. The imports came with all the settings fixed: they knew better than you did.

The Chaparral was an evolution. The lattice frame was lighter and stronger than the Scarab's. (Lattice was *Car and Driver's* word for the semi-space layout of four smallish steel tubes running fore and aft, triangulated and cross-braced.) The stressed firewall was magnesium, all practical places had aluminum instead of steel and because the basic structure was lighter the various other parts (shocks and suspension parts and brakes) could be smaller and lighter as well.

By no chance at all, weight distribution for the Chaparral was fifty-fifty static. The engine was well aft of the front wheels, in the mid-front position (in the middle of the car but in front of the driver). Hall had selected the position to achieve the low polar moment of inertia. This means simply that with the mass in the middle, the car won't rotate laterally—spin out, in other words—as easily as it will if you have the dumbbell effect, with the engine in front and the transmission, differential and fuel tank in back. It's also true that when the low-moment car gets loose it stays loose, but that's something the driver should be prepared to counter.

Troutman and Barnes were collecting the dividends from their learning curve, using what they'd been taught by their own special and the Scarab.

Hall was in the middle of his curve.

Car and Driver reported the new car on the basis of Troutman and Barnes doing the work and planning while Hall provided the backing.

It wasn't quite like that.

Hall had bought a design rather than a finished car, he said years after he'd retired from racing. "When I had a car I'd helped design, I couldn't bitch.... That's when I began applying my engineering training to racing cars. It took some time." Hall's early driving and preparation were done in cars designed and built by others, so their shortcomings were something he simply had to accept and work around. Now, though, when he'd been one of the men who made the mistakes, as it were, he could begin to judge the designs, and the men who made them.

Troutman and Barnes were experts, in their field. They could and did correct and improve on any of the structural problems that cropped up with Chaparral 1. What they couldn't do was be of much help sorting the car out, learning why it didn't handle as well as it was supposed to.

There was another long-range problem. Hall and Sharp were ostensibly amateurs, and they had one race every month or so. They'd race and by the finish have some idea of what the car was doing, and what they were doing. Then would come the layoff and they'd forget, lose their touch.

Little of this was apparent in the beginning. The Chaparral 1 was within shooting distance of the Riverside record during testing, and it got better. Yes, so did the others. But the team built two cars for Hall and Sharp, and Troutman and Barnes built three more, as they went along and as the orders came in. The customer cars had a list price of $16,500. That the Chaparral 1 was clearly the best front-engine special on the market didn't matter much, because first, there had been a ready supply of such racing cars for so long that everybody who wanted one had one and second, the configuration wasn't the way the world was moving.

The Chaparral 1 was second in its first race, and Hall and company got back to work. The car was in the hunt, second again at Riverside in 1962, winner of the Road America 500 that year. Two of the customer cars went to good customers; the Meister Brau team replaced the front-engine Scarabs, and got some more wins.

Even so... "the Chaparral One came at the wrong time, too late and there wasn't enough time to sort it out," Hall said later. "If it had begun with proper brakes and handling, it would have been better."

By 1962, though, it was clear that the Chaparral 1 wasn't going to be collecting top honors much longer, so "Hap [Sharp] and I decided to find a transaxle and put that Chevy in the back," said Hall.

Meanwhile, other builders and designers were at work. Some did well, but most didn't.

Dailu

Another Canadian special was the Dailu, so called because the first names of the builders were

The Dailu as it looked at Palm Springs, after it had become a vintage racer. This was a front-engine Canadian special with Chevrolet power, Jaguar suspension, a homemade frame and some handling problems. As the additions to the front panels indicate, the aerodynamics were a long way from correct.

Ol' Yaller III was as much neater than II as II was neater than the original—that is, a lot. It had Buick power beneath that aluminum bodywork. Road & Track

David and Luigi. They were based in Montreal and made their first car from parts that worked with other cars—a Corvette V-8 and gearbox, Jaguar front and rear suspension—with a homemade frame of square steel tubing. The first car was wrecked and burned in Nassau Speed Week in 1962, so they made two more front-engine versions and then a rear-engine special, still with Chevy and Jag parts, and finally rebuilt the first example but with so many new pieces that it was officially Dailu number five.

None of the cars had a top finish. They were able to stay with the smaller rear-engine racers on sheer power, but when the newer cars got better and bigger engines, the Dailus faded away.

Ol' Yallers III-IX

Bill Frick was the man who invented the Cad-powered Studebaker and Ford. He was also the pragmatic builder who helped Briggs Cunningham in those early years.

Ol' Yaller IV was longer than the earlier models and was supposed to be more of a road car, although it doesn't look like that here. Road & Track

In the thick of the action, Ol' Yaller III carries its famous Idaho license plate through Turn Nine at Laguna Seca.
Road & Track

Frick had a shop on Long Island, and displayed on the wall in the front of the shop was a sign that explained the meaning of life: All You Need Is The Money.

Well, all right, money's not the *total* meaning of life. Nevertheless, only a few people ordered Chaparral 1s at $16,500. A few more took the Cooper Monaco (the Climax-powered sports racer that had eclipsed both the front- and rear-engine Maseratis as the car to have late in 1961) at ten grand and change. But Max Balchowsky kept hearing from people who wanted one of his cars, which he would supply on order and to specifications for $4,500.

Not as many people as hoped for put their money where their mouths were. However, because of the presumed demand, Balchowsky and Cal Metal Shap-

Restored for vintage racing, Ol' Yaller IX sits in readiness at Palm Springs. Builder Max Balchowsky and owner-driver Ted Peterson have swapped and replaced so many parts that they don't always remember which car this front was originally made for. Still yellow, of course, but the rules require real racing tires and the wide whitewalls have gone.

ers made a body buck so new bodies could be semi-mass-produced in aluminum, and Troutman and Barnes rolled out seven bodies, so much better shaped than the first versions that it was almost hard to tell they were Ol' Yallers, at a cost of $2,000 each.

Ol' Yaller III arrived in time to be a viable racing car. It used the 401 Buick V-8, one of the steel tubing frames Balchowsky put together following chalk marks on the floor of his shop, and the usual mixture of Jaguar and Pontiac front end. The rear axles, live, usually came from Studebakers because that brand had a wider selection of gear ratios available.

There were some variations. Car VIII was based on a Jaguar center with homemade front and rear sections. Later cars had independent rear suspension from the Corvette Sting Ray, while before that if the customer wanted independent rear suspension Balchowsky used the center-section Salisbury made for the English Peerless, which a sharp bargainer could get for twenty dollars.

In its early days Ol' Yaller III was a contender. Bob Drake, who Ted Peterson says was much better than his record shows, was fifth in the 1961 Times Grand Prix. This was big-time, as the winner was Jack Brabham in a Cooper Monaco followed by Stirling Moss in a Lotus-Climax. The combination ran in 1962 but pushed so hard the car began to break down. Late that year, as the "Troutman-Barnes Chaparral" (to quote *Road & Track*) began to win, Ol' Yaller III appeared for sale in the classifieds: Make an offer.

Alas, the clock had run out on that sort of car. The last of the Ol' Yallers wasn't much better than the first. These cars were long, with a 91 in. wheelbase, and weighed 2,100 pounds so not even the might of a 425 Buick V-8 could keep them ahead of the smaller, more agile, more streamlined competition.

The cars were useful and went on to other things. Car IX—which also bore badge number VIII because they only made eight badges—was driven by Elvis Presley in a dreadful movie called *Spin-Out*. And then the car was wrecked making the movie *The Love Bug*. (It should come as no surprise that Balchowsky and the people who made cars for movies were friends; Balchowsky did have his shop in Hollywood after all.)

Ol' Yaller IX sat in the proverbial barn for thirteen years, until Peterson rebuilt it. The car now has the front from Ol' Yaller III obtained in another of those complicated deals racers make, which fit right on except that the aluminum front was so bashed they copied it in fiberglass and then later found a craftsman who could take a splash of the fiberglass and make a buck and shape a replacement aluminum front for the car.

Balchowsky owned some property in Idaho, which is how he always managed to have the cars registered in Idaho and to bear the Famous Potatoes slogan, which Ol' Yaller IX still carries in vintage races. The car is easily the best-loved machine in vintage racing, and now as then the organizers get jolted if the car whips the six-figure Ferraris, which it has done.

The cars may have been a touch behind the times in their weight and the use of drum brakes right up to the last example, but Balchowsky was ahead in his appreciation of tire size and he put the power on the ground with wide (for then) rear rims. Ol' Yaller IX ran races with 8½ in. rims in back.

When Peterson beat the purebreds at Laguna Seca, the owners complained to the organizer, who came up to Peterson saying, next year you must run 6 in. rims in back to be authentic. Peterson called in Balchowsky as a defense witness—vintage racing is full of politics, just like other racing and just like real life—and Balchowsky backed up Peterson. Forget it, said the organizer, don't tell me about that car, I have it all documented.

You can imagine how Balchowsky, who built the car to make mock of such people, felt about that.

Cheetah

The Chevrolet engine was still making waves in 1962. It was so good in relation to size and weight that the USAC proposed a stock-block, versus race-based, formula for single-seaters.

Road & Track's comment was that the cost of building a pure racing engine is so high that there must be some promise of stability before any sensible person would invest in making such a limited-production engine. Ten years was the time *Road & Track*'s John R. Bond estimated for breaking even with the project.

Beyond that, Bond wrote of the stock-block proposal, "The idea is fine except for the fact that it in effect hands the job over to the Chevrolet V-8." And it would have, although the B-O-P alloy V-8 and the little Ford were coming along. The Chevrolet engine was the production engine to have. The problem was that there was no suitable transaxle to put the engine back where experience was teaching that it should be.

The middle seemed like a fair compromise.

Bill Thomas was a builder with experience in such matters. He'd done a lot of work for Chevrolet's special effects people, the ads that had cars doing what looked like the impossible, and he had been fooling with the rear- (as in hanging aft of the rear wheels) engine Corvair. His belief in the V-8 engine and in the people working behind the scenes at Chevrolet convinced Thomas to produce, or at least build, a sports-racing car that he named the Cheetah.

Thomas was aided by skilled racing car designer Don Edmunds, who had begun as a driver and was Rookie of the Year at Indianapolis in 1957, but had decided to build cars instead. Edmunds specialized in suspension, and in sprint car construction.

The Cheetah was mostly a sprint car with coupe body. Edmunds began with the standard tube frame, of chromoly steel. We might call it the four-tube frame, with a series of cross-members joining the lower pair, and with triangulation between the upper and lower pairs. The sprint car part showed up with the elaborate roll cage, to protect the driver and stiffen the frame itself. This last was a touch adopted later by the stock car builders, especially in the classes requiring stock—pause for chuckle—frames and bodies.

The sprint car heritage also showed with the engine placement. It was so far back that there was no driveshaft proper, just the engine, clutch housing, transmission and differential. The rear end came from the Sting Ray. Edmunds used coil springs over tube shocks rather than the Corvette's transverse leaf spring, and made his own trailing arms to locate the rear wheels. Front suspension was upper and lower arms. Brakes were Chevrolet's stock car drums: there was a lot of reluctance to adopting disc brakes.

Thomas chose the coupe body because first, that was a better shape for aerodynamics and comfort at speed and second, he liked coupes better than roadsters. The doors were hinged at the top and swung up and down, like those on the Mercedes 300 SL coupes of the previous decade. There was little attempt at creature comfort, though, with a minimum of upholstery and paneling. The seat, not much more than a fiberglass shell, was more covered with fabric than actually padded.

The Cheetah began life with several strikes against it. The first three bodies were aluminum, which is easy to use when you need only a couple of examples but which wouldn't have been practical in production. The idea was to make a few, send them on the magazine circuit, stir up interest and then go into production with fiberglass bodies taken from a mold made off the aluminum.

Cheetah	
Wheelbase	90 in.
Length	na
Tread	58 in. front, 57 in. rear
Weight	1,500 lb. dry
Engine	Chevrolet V-8
Displacement	327 ci
Claimed power	390 bhp
Top speed	na
0-60 mph	na
Quarter-mile	na
60-0 mph	na

The Cheetah was supposed to be a racing car for the street, or a road car that could be raced. But by this time that wasn't practical, unless it was a production class car and that needed at least 100 examples. . . . You can't steal an egg unless there are some chickens and you don't have chickens unless you begin with an egg.

The car was shown to the press, which was allowed some brief time at the wheel but not allowed to turn loose with the clocks. The engine in the prototype was a modified version, with the 327 ci used by the factory then, but with a hot cam, a higher 11.25:1 compression ratio and the Rochester fuel injection in limited production for the hottest Corvette.

If the car weighed 1,500 pounds and *if* the engine had 390 bhp, the Cheetah would have been blazingly fast. The weight distribution was given as forty-five

Cheetahs were coupes because builder Bill Thomas liked closed cars and thought they were easier on the driver. The aluminum body was supposed to be a prototype for later fiberglass, but owing to GM politics and racing progress the project never got beyond a handful of examples.

percent front, fifty-five percent rear, which sounds reasonable in light of the engine being that far back, so traction should have been no problem.

The Cheetah was still another attempt to make a Cobra version of the Corvette: same engine and drivetrain but minus 1,000 pounds and all the frills. Not for the first or last time, it didn't work.

Jerry Titus, of *Sports Car Graphic*, was a journalist who became a racing driver, one small step at a time. He was always willing to push a car to its limit and he seems to have done that in his review—not a test, because no hard performance figures were taken—of the Cheetah. Titus described feeling out the car on its fresh tires, and experimenting with traction under power. He mentioned the car's power, and his hesitation to risk the lone body before it was used for the mold for the fiberglass. Reading backward between the lines, the car seems to have been a handful. The Cheetah ran some amateur races but never was in the hunt in the big events of the day.

Paul Van Valkenburgh, who was a Chevrolet engineer at that time and working with the performance teams inside the factory and out, wrote an excellent history of Chevrolet's racing effort, public and otherwise. He said three Cheetahs were made for racing with the alloy body and thirteen more were made in fiberglass for customers and the street before Thomas realized that it wasn't going to work and went back to building nonsports racing cars.

Speaking of sports or nonsports, in 1962 the FIA revised the rules controlling sports cars or prototypes or whatever one wished to call specials or any other form of racing car that could in theory be used on the public highways as well as raced. The new rules were tighter than those of 1954, which they replaced and which were used by most builders of sports-racing cars, from Pete Lovely and his Pooper down to Bill Sadler and the Comstock-Sadler. The 1962 rules demanded doors on both sides of the cars, for instance, and a real passenger seat as well as the reworking of various dimensions.

The USAC had five professional road races in 1962, enough so that it and the SCCA got into still another tussle. The SCCA had 13,000 members by this time and reckoned, surely correctly, that most of those members had joined for fun, for rallies and parties and racing in one's spare time, with little thought (or hope) of getting into racing as a career. So the SCCA retained the amateur rules.

The club went beyond that, though, and began threatening the licenses of the SCCA drivers running USAC events that hadn't been approved by the SCCA. But they worked it out, and backed off. Just why was never explained in detail: One hallmark of SCCA history has been that the directors settle their questions in the boardroom and make an official announcement and that's that.

The cure, or at least what they did to keep both the ability to run major races and the rules against having professional racing, was to make it wrong for the driver to accept prize or salary or any money for racing.

Spot the loopholes? First, it's the driver who couldn't take money. The car owner could. Many a wife or father or cousin became the owner of record, quickly. Corporations or companies could own cars and accept prize money, and that happened just as fast.

Second, the driver couldn't be paid for driving. He could be, oh, a consultant, and get paid for that. As writer Dick Van Der Feen put it, the car owner couldn't pay the driver to drive. But he could pay the driver to jump over the car owner's foot.

What we had was blatant hypocrisy. And everybody knew it. But because it worked, allowing a mix of good amateurs and homegrown professionals and imported stars from overseas who all raced hard and fair, nobody objected all that much.

What we also got was some new cars. One was the Cooper Monaco, which had a Climax in the back and was a sports version, to the letter of the law, of the Cooper Formula One car that was then about the best in its field.

Lotus replied with the 23, a replacement for the 19 (which was known as the Monte Carlo). The 23 was built to conform to Appendix J, the FIA's rules for sports cars, and had a full-width windshield, a place for luggage and so forth. There was an ongoing debate even then about thinly disguised racing cars dominating the sports car class.

Early in the 1962 season the races were dominated by Lotus, Ferrari, Cooper and mixtures like a

The Cheetah's engine was so far back that the radiator could be laid down and cooled by jamming air under the front of the car and extracting it from the vents in the hood.

Roger Penske again, in a modified Cooper Monaco. This was a later version, and it had the full windscreen and side panels and large door required as of 1962 rules. The first few times Penske appeared with the sponsor's decal aft of the front wheel, the SCCA made him tape over the "Telar."
Road & Track

Birdcage Maserati with Ferrari four-cylinder engine. Ol' Yallers III and IV were there, but weren't competitive, and the big names were in imports. Hall's Chaparral 1 was the lone exception.

Cunningham's Cooper-Buick

Pioneer hot rodder Briggs Cunningham took the next step. His team had gone from Jaguars to Listers to Maseratis, and they were winning but not as frequently as Cunningham had in mind.

He saw his next chance. The newest Cooper Monaco was driven in late 1961 by Bruce McLaren, a New Zealander who was a good driver and a clever builder.

After the races Cunningham bought the car, but not the 2.8 liter Climax engine. The engine was removed and returned to Cooper in England. The car was shipped to . . . Reventlow's shop in Los Angeles.

Cunningham and his resident wizard Alfred Momo had decided that both the engine and the chassis from their Maserati were no longer the best that could be done. They'd taken notes when Rodger Ward used an older Cooper Monaco combined with one of the new alloy V-8s from GM. The car had looked good before running into mechanical trouble. (Ward, who was an Indy star, always enjoyed the sports car crowd and made history when he beat the sporty set in a Formula Libre race with an Offy midget.)

Ward was no fool and had even obtained sponsorship from Kaiser Aluminum, what with the alloy engine being the powerplant of the future.

Reventlow was given the job of developing and installing the little B-O-P V-8, which was sold in Buicks, Oldsmobiles and Pontiacs but most often was listed in the program as a Buick engine.

The stock engine displaced 215 ci or 3.5 liters. It was best improved with a mild increase in bore, to make the engine 3.8 liters. RAI used four Weber carburetors, an exhaust system with the pipe lengths chosen to create pulsing and gain power at a specific rpm, a hot cam and the usual work to ports and combustion chambers, and got an unstressed 280 bhp from the little V-8.

There were some surprise benefits from the conversion to the alloy engine. First and most obvious, the 3.8 liter domestic engine was 20 pounds lighter than the 2.8 liter four. Second, because the V-8 had more, smaller power pulses for a given engine speed, there was less vibration and strain on the gearbox and the rest of the chassis, so RAI simply had to match up the new engine to the original transaxle, making the conversion much easier than such projects usually are.

The numbers hold two surprises. First, the Cooper was much heavier than anybody expected. Cooper used thick-wall steel tubing for the conventional space frame and—unlike Lotus, where Colin Chapman was famous for providing no more weight than he had to—Cooper provided extra beef, just in case. Second, the wheelbase was much longer than RAI expected, or than American builders were used to. Because the Cooper was so low, it was assumed to

Cooper-Buick	
Wheelbase	98.5 in.
Length	na
Tread	na
Weight	1,480 lb. curb
Engine	B-O-P alloy V-8
Displacement	232 ci
Claimed power	280 bhp
Top speed	na
0-60 mph	na
Quarter-mile	na
60-0 mph	na

have been smaller than it was. And as the record shows, the length and weight didn't hurt.

Beyond that the Cooper portion was conventional. It had a steel tube frame with aluminum skin that came off in sections for easy repair and maintenance, independent suspension front and rear by means of upper and lower A-arms, spring-shocks, and additional control and tuning with anti-roll bars. Steering was rack and pinion, brakes were disc. The engine was aft of the driver and ahead of the rear wheels, the transmission portion of the transaxle was aft of the rear wheels, and the fuel, oil and water were carried in the front of the car, to keep the balance close to equal.

The Cooper-Buick went on to have a successful career, driven by Hansgen and others. It never achieved the domination that Cunningham's earlier cars did, but then the competition was tougher, and anyway, it was a happy return for Cunningham, what with getting back to Buick power after all those years.

The last Scarab

Reventlow Automobiles didn't ordinarily do such conversions as the Cooper-Buick for Cunningham. The firm worked on Cunningham's car for a reason beyond the fee, and apparently the crew took careful notes.

There had been the overwhelming front-engine Scarabs; then the Formula One debacle; then a brief flirtation with Formula Libre and Continental, the single-seat classes for stock blocks and outmoded Grand Prix engines, respectively. The Grand Prix car was obsolete even for those races, and even while they were working with the Scarab engine the Cunningham crew spoke of perhaps using a Coventry Climax engine for the free formula races the car was built for.

If timing is everything, then poor timing is worse than none at all. In the Australian Grand Prix, the only race in which the Scarab rear-engine single-seater raced, the car used the Buick engine. It was a good car, perhaps capable of winning races in that

Cunningham's team Cooper needed the bigger Buick V-8 to compete with the other imports powered by Detroit. By this time all the components for such an upgrade were readily located. Petersen Publishing Company

formula; the formula meanwhile didn't attract fans or entries or sponsors and died an early death.

Reventlow commissioned one last car. It was a logical and logistically inspired mix, a rear-engine sports racer. By that time, RAI knew how to build a good single-seater. It had constructed and in effect designed a good conversion, the Cooper-Buick for Cunningham, and by happy chance was busy with a Buick engine for the road racing single-seater at the same time that the legendary Mickey Thompson was working on his revolutionary Indy car, with Buick power. Both teams were getting subrosa advice from certain outfits not far from Flint, Michigan, and because they weren't in direct competition, they swapped discoveries and parts with each other.

This engine project was a challenge. Ward's earlier Cooper-Buick had showed promise but been plagued by overheating.

Just as with the fabled flathead Fords, the old-time hot rodders sat down and puzzled it out and found the cures. Thompson's crew had it easier because they got to burn methanol, which has greater flow and greater capacity for absorbing and carrying away heat (refer back to the Offenhauser experiments). The water passages were revised and the flow improved, and the oil system was enlarged to carry away heat as well.

Just why heat was a problem on the engine wasn't explained. Perhaps it was simply that the little Buick, advertised at 190 bhp stock, developed 155 bhp on the dynamometer when new, and was taken to 300 bhp by the end of the development. Twice as much power equals twice as much heat; the factory hadn't planned on the engine being used like that.

The next step was radical for the day and involved a flow bench. Most people now know that engine developers have devices with which they measure how much air will flow through a given shape in a given time at a given pressure. Then they change the

Scarab-Buick	
Wheelbase	91 in.
Length	na
Tread	52 in. front, 50 in. rear
Weight	1,200 lb.
Engine	Buick alloy V-8
Displacement	239 ci
Claimed power	250-300 bhp
Top speed	na
0-60 mph	na
Quarter-mile	na
60-0 mph	na

size and shape and measure again, and after only months of detail and infinite patience the engine is putting out the power needed. Porting and shaping the combustion chambers was the last of the engine builder's various arts to become accepted and practiced. It is how, long after hot cams, big carburetors and tuned exhausts, the people in stock and sports car racing manage to crank more and better power out of the same basic engines year after year.

This is now and that was then. *Sports Car Graphic* spent much of the Scarab's space explaining just how the flow bench worked, while readers lapped it up.

The Scarab's Buick was developed in stages, from 215 ci when new through overbore and a mildly longer stroke out to 239 ci. RAI used side-draft dual-throat Weber carburetors in a variety of venturi sizes matched to the displacement and cam timing and rpm. The 215 ci version logged 245 bhp with 40 mm venturis, but those venturis were too small for the 239 ci engine so one got bigger carb bodies and 47 mm venturis and reached a maximum of 300 bhp, according to *Sports Car Graphic*.

Road & Track's story said that RAI had seen more than 300 bhp on Webers, which gave more power than Hilborn fuel injection, but that the actual

The last Scarab had a different look, with a distinct wedge shape and peaked front fenders, for extraction of heated air from beneath the car. Note the low wraparound screen and what looks like a scoop or dam below the radiator intake.
Road & Track

race engine produced 250 bhp. *Sports Car Graphic* said the race engine was less than optimum. Probably both magazines were saying the same thing in different ways.

By this time so many racing cars were being built with powerful engines in the back that there was a transaxle just for such an application, the Italian Colotti. RAI used one, a five-speed, built just for RAI— it's nice when you can afford the best. It came with a quick-change rear for swapping final drive ratios, and with its own pressure-delivered oil system for cooling and reliability.

Dick Troutman and Tom Barnes had gone their own ways when the sports and formula programs were finished and abandoned. Phil Remington recalled that the design of the back-motor racer was handled by Eddie Miller, a designer and builder whose best-known work was a record-breaking streamliner that used Pontiac power... back when Pontiacs came with inline side-valve sixes.

Mechanically the rear-engine Scarab was conventional, with Buick V-8 behind the driver and transaxle behind the engine, a sturdy tubing framework holding all the parts in place and saddle fuel tanks on each side of the cockpit. Road & Track

In its later, used-car stage, the Scarab had a Chevrolet for an engine, Mecom for an owner and A. J. Foyt for a driver, and it won at least its share of races, as here, at Nassau in 1963. Road & Track

The magazines said the Scarab was by Lotus out of Cooper, and that was mostly the case except that the pattern for the frame came from the Scarab single-seater. Frame tubes were mild steel, triangulated for strength but with much of the rigidity coming from the size of the tubing itself. This meant more weight for strength than with the Lotus, which used a greater number of smaller tubes, but it went along with Cooper's practice, which was to give more weight and make it easier to repair if the need arose—as of course it does in racing. For the same reason the rear frame tubes that carried the engine unbolted so that engine and transmission could come out in one unit.

Front suspension was independent, with A-arms. The suspension proper came from coil-over shocks, with the shock bodies equipped to be adjusted in place, and with torsion bar anti-roll bars giving spring and roll rate adjustment. The coil spring on the shocks could be moved up and down, so the car could be tuned for ride height and even weight-jacked. Weight-jacking is varying loads from side to side and front to back to balance the car in two dimensions. That's another stock car trick, and one the road racing crowd adapted early. They saw how Kurtis and Balchowsky made their cars work and they cheerfully went with what worked.

Steering was rack and pinion and brakes were Girling disc. RAI had the rear hubs and the wheels cast in magnesium, the best balance of light versus strong with cost no object. The wheels were 15 in. in diameter and the brake discs were 11 in. The rims were wide, as that useful dimension was becoming public knowledge, with 5.00 section tires in front and 6.50 rear. The new hubs were needed because the rear suspension was reversed A-arms and long trailing radius rods, with upper and lower links widely separated for geometric reasons: the distances and angles are critical here so the wheels will tilt, or not, as they go up and down.

Surprisingly the car wasn't given detailed examination or inspection by the press, so we don't have the full set of figures we might like. Wheelbase was 91 in.; tread was 52 in. front, 50 in. rear; and the car stood 32 in. high. The engine was nominally Buick, with between 250 bhp and 300 bph. If we treat the Scarab as a cross between the single-seater at 1,000 pounds and the Cooper Monaco with Buick engine at 1,400 pounds, then the car weighed 1,200 pounds dry.

The body was aluminum, a lovely shape that also met the FIA's rules to the letter. Reventlow went one step beyond that, as he registered and equipped the car for road use in California. He left out padding and trim but had both seats accessible and upholstered in gray vinyl. The interior panels were polished alloy or painted in black crackle. The steering wheel rim was hand-shaped and polished.

This last car was extra-special. Reventlow was a car nut, in his heart and not just for social or ego reasons. The third of the first series of sports cars was redone for highway use after it was outmoded for racing, and Reventlow presumably planned to do the same thing here, except that because there was just the one car, it had to be done for both purposes.

The car was raced, and did reasonably well in Reventlow's hands. Then it was sold to Texan John Mecom, who was in the process of assembling a team of sports racers, and when driven by A. J. Foyt racked up at least one major win.

The last Scarab was as good a piece of work as the first ones had been. So why did they stop?

One reason was taxes. RAI had been in business five years. The ostensible purpose when the builders began was to build cars, win races, build a reputation and then go into production and get the money back

The sprinter that never was

Can't prove a negative, eh? Consider the sports car Bill Stroppe *didn't* build in 1961.

Stroppe was the pioneer hot rodder and builder who helped make the reputation of the Kurtis sports car back in the early fifties. He was a good, hard driver, but because driving was a talent shared with many racers and organizational ability was rare, he became a team manager and builder and creator of projects for people like Ford and Lincoln. Driving was a hobby; running teams was how the bills got paid.

But Stroppe was still interested in fun, and in theory. One of his more talented employees was Don "Red" Edmunds, technical wizard and also a some-time driver (Rookie of the Year at Indianapolis in 1957).

Edmunds had some ideas about sports cars. While working for Stroppe, he wangled the boss' permission to design and build a sports car along sprint car lines, much as Kurtis had done a decade earlier.

The pictures shown here are all that remains of the project.

The frame was four good stout tubes running parallel and doing the job the old ladder-type frame used to do.

The engine was offset to the left, by at least six inches. Stroppe recalled that this was thought to be one way to balance the driver's weight on the right and keep the car in perfect trim, no matter which way the turn went. (As seen earlier, as many builders worked to bias the car to the right, for an advantage in the right turns that predominate in road racing.)

The car was front engine, with the engine in the middle and the driver right back against the rear axle. This was how it was done when Edmunds began thinking about the car, but by the early sixties

Bill Stroppe's never-completed sports racer was strongly influenced by sprint car practice; witness the live front axle with torsion bar, the engine offset to the left and the driver on the right. Also note the center, vertical arm on the axle. That's part of the Watt link keeping the axle in position laterally. Road & Track

the engine was known to work better at the back of the bus unless, as with Jim Hall's first car, there were other advantages or strengths.

Next and most incredible, the Watt link suspension was used in multiples. Each axle had a Watt link in its center, to locate the axle side to side while the frame could tilt under cornering loads as the axle stayed horizontal and the wheels vertical. Each wheel was a fore-and-aft Watt link so the axles couldn't go back and forth but could allow the lean required.

Damping and control were provided by the coil shocks on each side, and suspension came from the torsion bars at the extreme ends of the frame, with trailing (or leading) links joining the bar with the axle.

The car had sprint car steering gear, steering wheel position and seat. The differential was Halibrand quick-change with axles and housings modified to allow the engine offset.

The engine was from Ford, which had introduced a much hotter V-8 and some performance options for the new compact cars, and where the Mustang was a secret project. Stroppe had close connections with Ford and had been one of the first to obtain and develop the new powerplant for racing.

The aluminum body pieces were pounded and shaped over the body buck, a wooden form. When the pieces were a perfect match to each other and the buck, they were welded and shaped until the seams no longer showed. The style of the car was contemporary and a good match for, say, the Chaparral and the Zerex specials.

Pictures of the car were taken by Lester Nehamkin, for *Road & Track*. They went into the library files instead of the magazine because Stroppe never completed the car. Financial reverses, he recalled, made the team suspend everything that wasn't going to help pay the bills, and the sports car had no sponsor.

Although Edmunds was current when he began, by the time the project was this far it was obvious that the car would be too heavy, and would be handicapped by the engine location, even if the Ford V-8

Here's the live rear axle, the upright seat and a much clearer look at a Watt link. The vertical arm is just to the right of the quick-change differential, with the upper horizontal arm going from the top of the vertical link to the frame rail on the right. The lower horizontal link goes to the lower left frame rail. Road & Track

139

The body was shaped on a wooden buck and the various pieces were joined with seams that were later filled in and blended into the panels. This would have been a lovely car; too bad it never got beyond this stage. Road & Track

performed—it certainly did, by the way—or if the suspension worked on the road courses of the day. One is tempted to suspect that the suspension wouldn't have worked; in the coal-cart era of leaf spring and live axle, all the systems worked about the same because none of them allowed more than an inch or two of wheel travel. This was before Colin Chapman and rivals discovered that if you allowed the wheels to move, in fact allowed the body and frame to move up and down, the tires were on the ground more of the time and could keep the car on the track and moving forward. Longer travel meant the torsion bars, with their links and levers, were more difficult to locate because the links had to be long so the wheels could travel.

We'll never really know. Stroppe had forgotten all about the project when asked about it twenty-six—or was it twenty-seven—years later. Face to face with the photos he remembered the car but could only vaguely piece together the financial problems that led to the sports car project being suspended and the car itself sold for a song.

The sprinter was never seen again, as far as Stroppe or the racing record can determine. Seems a shame, because the workmanship and design were beautiful.

selling cars. Or so they told their accountants. The dream is all very well but the Internal Revenue Service doesn't dream. It has a rule that if you don't make money after a certain point, you don't have a business, and you can't deduct the money spent.

Remington said there was another factor. Reventlow had married a movie star who was known in racing circles for hoping her husband would get this foolishness out of the way so they could go back to Hollywood. The wife was a big spender, Remington said, and went through as much money as RAI did.

The bottom line was that with the personal pressure on one hand and the Internal Revenue Service on the other and with his having proved several points about his talent and ability behind the wheel and the desk both, Reventlow quit racing and sold everything except the road-equipped first sports car. And that was that: Reventlow was the last sportsman willing to put his neck and his bankroll where his enthusiasm was.

Zerex Special

Roger Penske was a lot like Reventlow, while at the same time he was completely different . . . and much better suited to his times.

Penske came from a successful upper-class family and he came with brains and drive and ambition and guts. Only the brains were visible at first. Early in 1958 Penske turned up at an SCCA driver's school, equipped with a membership card and a fuel-injected Corvette. He'd picked the hottest production car on the market he said later, because it seemed to him that the object of the exercise was to go fast so it made no sense to begin going fast in a slow car when he could use a fast one.

Penske was good enough to earn his racing license at that first school. He got his national license, in fact, and three weeks later led his class at an SCCA national until the car overheated.

Some of this prodigious talent was misleading. Penske began motor sports on a motorcycle, which he street raced until a crash with a car put one leg in a cast. He had MGs, went drag racing with domestic iron, bought an XK120 Jaguar and raced in solo events on dirt ovals of all places. He was competitive, and had resources of his own and a way of getting other people to help.

Penske also had a knack of being in the right place at the right time, or perhaps of creating opportunity. His list of racing cars after the Corvette sounds like a list of good cars of the day: the Porsche RS and RSK, Jim Hall's old RS-60, the Tipo 61 Birdcage Maserati; then one of the first three Cooper Monacos in the world. Along the way he'd won two SCCA national championships and been named Driver of the Year by the *New York Times*, and he bought a Cooper Formula One car and was eighth in the US Grand Prix at Watkins Glen in 1961.

Driving was only part of the story. As Penske told *Road & Track* late in 1961, a talented driver needs to work hard and find some way to pay for his racing, which in 1961, when dollars were much larger than they are now, added up to about $10,000 annually.

It's the start of the big race at Laguna Seca, 1962. The man in the lead and in the middle of his car is Roger Penske. That's Jim Hall in the Chaparral right behind, and there's a Cooper Monaco at far left, a Scarab two rows behind Hall and another Scarab just passing the starter... no matter. The little Zerex Special is obviously one jump ahead. Road & Track

Penske had friends, which was an old story.

And he had a sponsor, which was a new story. We've seen the Comstock-Sadler, and Rodger Ward's Cooper-Buick sponsored by Kaiser Aluminum, so the idea was out there.

Penske had graduated college, married and gone to work as a salesman for Alcoa while he was winning those races and titles and awards. He knew how to promote, and how to sell.

Penske went to DuPont, which had just introduced a coolant called Telar. He made a presentation based on the now-accepted idea that if DuPont sponsored his car, he'd call it the Telar Special and the name would be on the car and thus in front of the public, to make the name known and convince the fans, who almost certainly had cars of their own and probably did at least some of their own maintenance, that Telar and DuPont were names to know.

It worked. Penske got the contract and the sponsorship. Some noses were out of joint early in the program, as the SCCA made Penske tape over the Telar part of the emblem, but DuPont kept on, and so did Penske and he won the USAC national championship in 1962.

The stage was set.

When Penske told *Road & Track* that he was in racing for fun and didn't plan to make a career or business out of it, he was on the Alcoa payroll (and large corporations seldom enjoy hearing that their employees have plans of their own). At about the

The Zerex Cooper was in profile just what it was supposed to be, a Formula One car with enclosed wheels. Road & Track

same time he visited Cunningham's shop. He happened across the Grand Prix Cooper that Bruce McLaren had crashed in the US Grand Prix. Like Hall, Penske knew that he didn't have any chance of getting the newest and best cars from the racing factories. This was a new car, but it was bent so badly that most people saw only some spare parts.

Penske saw his chance. Road racing was an extended family and everybody knew everybody else. Penske had a Maserati and parts and knew Cunningham was always willing to swap, so he gave his parts plus boot for the wrecked Grand Prix Cooper.

He and mechanic Roy Gane and some other friends hauled the wreck back to Pennsylvania. There they stripped everything; cut away the bent and broken sections of the frame; cut away the suspension mounting points that had been ripped, bent and twisted; and figured what to do next.

What they figured to do mostly was convert the Formula One car into a sports car, a sports racer, one of the best and best-known thinly disguised racers to ever pose as highway transportation.

Penske later estimated that they cut away and junked thirty percent of the frame. They built a jig and put the rest of the frame in that and replaced the missing sections and redid the suspension mounting points and the suspension and the wheels and so forth. At that point Penske had an up-to-date Grand Prix Cooper, a single-seater. It had a five-speed gearbox, disc brakes, a fully independent suspension and a 2.7 liter Coventry Climax engine in full race trim. Call it 1,200 pounds with 250 bhp. The Cooper could not have been bought in one piece. Penske had in fact ordered such a car, and been promised one by John Cooper, but it never arrived; the European factories liked the way their customers raced, but customers were all the Yanks were to them.

Penske then decided on a way to have an advantage, fair or not depends on how one interprets the rules. His first step was to call Henry Banks, director of racing at USAC, and explain his plans. Banks said fine, go ahead, so Penske enlisted Bob Webb, an Indianapolis craftsman who did the panels, and Larry Tidmarsh and Roy Gane—Penske has always been careful to give credit where due—and they made a sports car body.

Sort of. They made a tiny aluminum chassis cover, with bulges for the wheels and scoops for the brakes and oil and water radiators and little bitty lights tucked in the grille. There was a passenger seat, sort of, off to one side and the driver sat smack in the middle, just as when the chassis was covered by a Grand Prix body.

This occurred early in 1962. Just that year the FIA had revised the rules for sports cars or prototypes, and made them more normal in that the seats were supposed to be one on each side of the car's centerline and available for occupancy when the car was being raced.

This was a different sort of special, as it was built for one class and converted at home to run another class, but with the same engine, suspension and so forth as it had when new. Penske showed up with this bold new idea for the Times Grand Prix at Riverside.

Meanwhile, the SCCA had done another revision and had come up with the United States Road Racing Championship (USRRC). The USRRC had two championships: a manufacturer's title for Grand Touring cars, and a driver's title for sports racers/prototypes. The sports cars had two classes, one for engines displacing less than two liters, the other for engines with more than two liters. This made it easier to watch the race, and tell who was important, and thus easier to promote the events.

The USRRC was done in conjunction with the FIA. All the events would be open, so professional drivers from other clubs and other countries could compete, while amateur drivers who wished to retain amateur status could do that and race. It had been a long process and a delicate series of compromises and political moves, but the SCCA finally had a pro class and could keep the amateur program going at the same time, and would—although surely this wasn't part of the plan—outlast the USAC and become the only major road and sports car racing club in the United States.

In the first USRRC event, under FIA rules, a big flap arose over whether the race would be run for cars that met the 1962 FIA rules, or cars that met the 1954 FIA rules under which the SCCA had been running ever since 1954. We were big enough to write our own rules, as John R. Bond wrote, and we did like to provide a grandfather clause so racers with older machines weren't outlawed unfairly.

The Zerex Cooper versus Tim Mayer's sports Cooper makes it plain how much more compact the converted Grand Prix car was. Road & Track

That worked just fine. As late as 1963 Ol' Yallers III, IV and VII were still turning up in national races and even the HWM-Chevrolet, the first effective transplant of the Corvette V-8, was being raced seven years after it was built. Even so, rules are rules and gentlemen are gentlemen and sometimes those two clichés collide. They did at Riverside.

The rules had been ironed out by then, or so people thought. Jim Hall won his class at Sebring and Dan Gurney won the Daytona Continental for Lotus, with Hall and the front-motor Chaparral 1 in third. Jim Hall and Hap Sharp won the Road America 500 with the Chaparral 1 and Pedro Rodriguez took the Bridgehampton 400 with a four-liter Ferrari. Good close racing.

But when Penske arrived with his Cooper, identified strongly and consistently as the Zerex Special, nobody knew what to do. The rules were clearly

This is the later version of the Zerex Cooper, driven by Hap Sharp at Pensacola, Florida. By this time Roger Penske had moved the seat and provided the full windscreen, as the FIA dictated. Road & Track

against having the one seat in the middle and the second out of reach. But Penske had checked with SCCA and USAC officials before building the car that way, and they had said they saw no problem and Penske had proceeded on good faith.

He came.

He qualified.

He won.

It wasn't all that easy. By agreement, Penske moved the driver's seat half an inch to one side, so the seat wasn't actually in the middle of the car. The stewards of the meeting then voted to accept the car despite the uproar.

In the race, against no fewer than fifteen Lotus 19s, seven Porsches, nine Maseratis, three Chaparrals, two Scarabs and one Ferrari, and against drivers like Reventlow who spun his rear-engine Scarab, and Ward who put his Chaparral over an embankment, and the world-class Bonnier, Hill, Gurney and Hansgen, Penske was smooth and cool and did everything right. It was a victory, *Road & Track* said somewhat sourly, "darkened by his *hardly* disguised [as opposed to *thinly* disguised] Formula One car."

Next, the Zerex-Duralite Special retired from the Northwest Grand Prix, won by Gurney in a Lotus 19. Penske was second in both heats of the Pacific Grand Prix at Laguna Seca, and was awarded first overall on points. According to *Road & Track*, "Neither Hansgen's Cooper-Buick nor Jerry Grant's Lotus-Buick are quite getting up to the job. Jim Hall's Chaparral being the only disputant to their superiority, the European brands still have the best of it."

Well, perhaps. More in balance would be the view that Penske's restored Grand Prix car was better than anybody else's sports car. Penske gave credit and said that the Climax engine as delivered to customers was famous for breaking. His two Climax fours, though, were dismantled and rebuilt and carefully checked out by Roy Gane, and gave no trouble.

The entry lists show that the smaller, rear-engine cars from Germany and England were beating the larger, front-engine cars from the United States, England and Italy and that everybody was moving toward the middle, as in the new Scarab, the rear-engine Ferraris and Maseratis and the B-O-P-British hybrids.

Penske and the Zerex Special won the Puerto Rican Grand Prix, last in the 1962 series, with Tim Mayer second in the former Penske Cooper Monaco. The race was run under the SCCA rules for modified sports cars, which, *Road & Track* sniffed, "Penske's car meets the letter of."

To create an anti-climax, Penske converted the Zerex Special to meet the 1962 FIA rules, providing a windshield and luggage space and two seats of the same size and so on, but the old Chaparral won the race for sports cars and Innes Ireland in the ubiquitous Lotus-Climax won the main event. So there.

Except that Penske replaced his SCCA national championships with the USAC national championship for road racing. And he won $21,000 in prize money during his hot streak.

Putting that in context, when he was interviewed in 1962, Penske said that his annual efforts as a subsidized amateur cost $10,000, including the car. He didn't say how much the former Grand Prix Cooper cost to buy and convert, nor did he say how much he got from DuPont. We can guess, though, that he made a profit on the operation.

More important in the long run, Roger Penske proved to DuPont and the world that sponsorship was a valuable tool that worked in both directions.

Penske had several last words in this matter. He told the press, with a skill at which he has few equals, that the SCCA had always allowed machines such as his, that Americans had a tradition of building specials (he cited those two popular extremes, Balchowsky and Reventlow) and that other cars such as Pete Lovely's Porsche-Cooper and the Genie-Corvair and the production Cooper 1100 Sports had seats in the center and ran without anybody getting upset.

What he didn't say, surely because he was smart enough to not rub it in, was that to some degree the furor was because he had proved once more that the people controlled by a given set of rules are more determined and intelligent and visionary than the people who write or enforce the rules.

Having made his point and his reputation, and probably having gotten as much prize money and publicity and speed out of the car as there was to be gotten, Penske sold the Zerex Special to Bruce McLaren, signed on to drive for Texan John Mecom's team and shipped his Cooper Monaco off to Troutman and Barnes for installation of, yes, a Chevy V-8.

Cooper-Chevrolet

Back in 1957 the Automobile Manufacturer's Association imposed on itself a ban on racing involvement—in theory. In fact, most of the members didn't take part in the new sports of stock car and sports car racing and they seemingly saw the ban as an easy way to make points with the safety establishment.

Be that as it may be, the managements of Ford and Chevrolet, at the divisional level at least, got out of racing visibly and found other outlets quietly. OK, make that secretly. (There's also lots of evidence that the people at the top were careful not to know about anything they'd have to forbid, if you can read between those lines.)

There came into being an experimental task force within Chevrolet, some of it allied with the Corvette design and engineering group, some of it in the sedan section. They farmed out interesting projects, such as finding out how much power an outside developer could get from a stock-block Chevy engine, and they did other work inside. There had been a racing

Corvette program and now there appeared the lightweight Corvette Grand Sports that went racing under private banners.

The various GM divisions were fierce rivals, combatants even despite having the same corporate owner. They had separate engineering staffs and so battled with real weapons. The B-O-P alloy V-8 was sold in different form for Buick, Olds and Pontiac. It turned out that the Olds version had cylinder heads that flowed more and better air than the Buick heads did, so there were specials with Buick power and specials with Olds power and a couple with Buick engines wearing Olds heads.

One of the sports car magazines wrote that the emergence of the B-O-P engine meant the end of Chevrolet's reign.

Not if Chevrolet's racing group could help it. And it could, and did. One of its weapons was an alloy block for the Chevy V-8. Make that the small-block Chevy V-8, because what had been the division's truck engine was becoming available for passenger cars. It wasn't yet a performance engine, though, in part because the chassis and tire combinations of 1962 had more than they could do transmitting the power from the 283/327 ci V-8.

Independent testing had established that the closer the tuners made the Buick and Olds V-8s to the Chevy V-8, the better the B-O-P worked. The advantage of the newer engines was the lighter weight and better heat transfer of the alloy block and heads, so what the Chevy racing-experimental group did was design and begin production, much limited and only to the right people, of a small-block Chevy V-8 with aluminum block and heads.

It was some project, and it's hard to imagine how it came out of Chevrolet Division or General Motors

Penske's Cooper-Chevrolet began life as a Cooper Monaco but with larger and stronger frame tubes and related hardware. Coopers were always built strong, so the extras provided a margin and the conversion was relatively simple. Petersen Publishing Company

Bodywork for the production Cooper sports racer was more encompassing and maybe not as attractive. But the car was reliable and predictable and worked fine with Chevrolet power. This example was sponsored by Nickey Chevrolet, a Chicago dealership with strong ties to the racing people at the factory.

unless there was some careful turning of blind eyes by the heads at the top.

Roger Penske had friends at the top, and in the shop. By this time he was high enough himself so he didn't have to get his own hands dirty—although both Penske and Jim Hall were perfectly willing to do so, and to do any job in the shop, when needed. Penske got hold of the new aluminum engine and a Cooper chassis, except that the frame tubes for the Cooper were heavier, and thus stronger, than the tubes of a Cooper Monaco built for the four-cylinder Climax engine.

The engine and frame, suspension body, in fact the complete car minus standard engine, went to Troutman and Barnes, where they already knew a lot about the car and the engine.

The conversion wasn't difficult. The frame was a space frame, with four widely spaced longitudinal tubes braced and crossed front, center and rear. The rear of the frame was modified to clear and carry the larger V-8. The radiator that used to hold oil and water in separate sections was changed to carry just the water, with oil coolers elsewhere. And the fuel capacity was enlarged with saddle tanks, one each side of the cockpit.

The accounts of this car are vague when it comes to facts and figures. This is because although the Cooper-Chevrolet was bound to be a better car than the previous Zerex Special, it wasn't as much better than the opposition. Everybody had figured things out. It wasn't just big old front-engine cars against new little rear-drive cars, it was full fields of specials and factory cars with rear engines, fully independent suspension, disc brakes . . . and big V-8s. Not all from the United States, but from Europe as well, and in that context more than a few teams were busy putting Italian engines into English cars.

This meant that Penske's Cooper-Chevrolet wasn't the world beater his converted Grand Prix Cooper had been. So all the magazines did was make pre-race guesses that the engine surely had 300 or so bhp, while the car itself would weigh only a few pounds more than stock: the weight came from the added power, which put more heat into the cooling system and more strain into the frame and suspension and brakes, all of which had to be a bit bigger than before.

As John Christy wrote in *Sports Car Graphic*, "Yes, the big howlers are back in force."

Indeed they were.

Genie-BMC

The progression to bigger cars can be neatly traced with the work of one special builder, Joe Huffaker, and his semisponsored efforts with British Motor Corporation (BMC). Huffaker was visible earlier when he made several competitive cars with the help of the San Francisco-based importer, BMC. The team turned for a few years to building Formula Juniors, the single-seaters raced under a set of rules that was supposed to keep prices down. The rules didn't limit costs, nor have such plans ever worked, but Huffaker's company Genie (as in the magic creatures that make wishes come true) did good work in the class until it was obvious the class wasn't going to survive.

Huffaker could clearly see what was going to happen next, so beginning in 1961 he designed a small sports racer to compete in G Modified, which was the next-to-smallest SCCA class and limited to 1100 cc engines.

The design began with a true space frame in a Birdcage style, meaning that the network of tiny tubes filled the space within the body envelope and created strength through numbers and position rather than weight. The Genie frame was in three main sections: the front, with suspension, steering and radiator; the cockpit; and the rear, with engine and drivetrain. The front and center sections were further stiffened with stressed panels of aluminum and the engine bay included a bolt-in brace so it could be fully triangulated.

The suspension was independent, as with almost all specials built since 1960. Huffaker fabricated his own arms and links and bracelets and placed the pickup points and mounts so they were widely spaced, spreading the load. Coil spring-shocks were at each corner. Steering was the ubiquitous Morris Minor but with extra weight trimmed off, and brakes were Lockheed disc, 9 in. in diameter. (Disc brakes were virtually universal by then, especially for the smaller cars, and were in such demand that there were rival suppliers—Lockheed versus Girling, for instance.)

Weight was saved everywhere. The steering wheel came from a go-kart and the handbrake was simply a lever arm that pushed down on the brake pedal. (The rules said there had to be a handbrake; just how much braking it did was less important.)

Engines had come a long way since the Crosley and Saab days. The mass-production factories of the

Genie-BMC	
Wheelbase	88.5 in.
Length	na
Tread	50 in. front, 49 in. rear
Weight	726 lb. dry
Engine	BMC A-series
Displacement	67 ci
Claimed power	95 bhp
Top speed	na
0-60 mph	na
Quarter-mile	na
60-0 mph	na

world had done something along the lines pioneered by Detroit after World War II and had invested in research and modernization. Their products were much smaller and had fewer cylinders (four versus eight), but they were overhead-valve and sometimes overhead-cam and they were engineered to take more stress than their states of tune could give them.

The Coventry Climax racing engines enjoyed a short monopoly but British Leyland (MG, Austin and other captives) and Ford of England had good little sedan engines in class, 950 cc and up, and the tuners worked until the former dull engines—cooking en-

The Genie was built to FIA rules, and to the principle that one basic design could expand or contract to suit different engines and classes. Road & Track

gines, as the English say—were putting out as much power for as long as the racing-only engines did.

In the Genie's case, Huffaker being allied with the importer of British Leyland brands, the engine of choice was that from the small MG and small Morris, the BMC A-series engine. This came stock with 950 cc, approximately 58 ci, but to meet class G was bored out to 1096 cc, 67 ci.

After the usual reworking of ports and valve seats and valvetrain, the use of a hot cam, a compression ratio of 12.8:1 and breathing through a single 45 mm Weber side-draft carburetor, the Genie was propelled by 95 bhp. Clever machine work adapted the engine, which came from the front of the production car and sat in the back of the Genie, to a transaxle from a Volkswagen, which of course had the engine aft of the gearbox and driving wheels. But the gears were sturdy and the differential already had the fittings for independent suspension and the gear ratios were changeable because Volkswagen had used the gearbox for sedans and wagons and trucks. The Volkswagen transaxle came in handy.

The Genie's body was more worked than styled. The dimensions and fittings followed FIA dictates as to windshield and such, while the required lights were as small as they could be and still have bulbs. In profile the car was a homely miniature Lotus. The center stressed panels were aluminum, and the front and rear pieces were fiberglass. Huffaker planned the car as a limited-production racer for sale, so fiberglass made the best and most efficient body.

Journalism has its stages, its ebbs and flows, just like any other art. When the sport of road racing was just beginning the magazines had the time and the space and the cooperation of the owners and teams to take the car to a track and use instruments and truly

The space frame deserved its nickname of birdcage. The Genie's alloy panels stiffened the steel tubing frame and directed airflow as well. This was the smallest model, with BMC four power. Road & Track

The BMC-powered Genie could use a transaxle adapted from the Volkswagen Beetle, and neatly fabricated rear suspension from the builder. Road & Track

Laguna Seca before buildings: Jerry Grant in a Buick-powered Lotus 19 gets the hole shot on Chuck Parsons in a Lotus 23 and Dave Ridenour (hidden by the starter) in a Genie. Road & Track

The Genie-Corvair was a natural, as a combination and as a place for an accessory maker (EMPI) and a dealer (Friendly Chevrolet) to share sponsorship. Petersen Publishing Company

test, report and evaluate. Then, when the sport expanded and there was more for magazines to do, the track test went into eclipse. At the same time the staff members tended to enjoy themselves rather more than required, so we readers got more impressions and tall tales than we really needed.

Genie-Corvair

When imported cars began to take a share of the domestic market, the American manufacturers replied with the sensible introduction of made-in-the-USA versions of what the imports offered—that is, smaller and more utilitarian cars.

General Motors noted that Volkswagen had a distinctly different car: air-cooled engine behind the rear wheels, pancake-style four-cylinder engine, body shaped like that of an experimental car from the thirties. So Chevrolet did an enlarged version of the Volkswagen with a pancake six aft of the cockpit and driving the rear wheels, air-cooled with a fan atop the pancake, and with a very different style to the body.

The good news was that the Corvair did appeal to the sort of buyer who liked something different and inspired by Europe. The bad news was that there weren't all that many buyers with a taste for the unconventional.

The worst news was that the Corvair had a lot of problems. It wasn't a bad car but it was different. Its differences combined with inept public relations from GM changed the course of automotive (and legal) history, but that is another story for another time.

In this instance, the Volkswagen had attracted loyal fans who liked to improve their cars, and the Corvair did the same thing. Several firms, EMPI for one, catered to the Volkswagen market and also expanded into Corvair parts and equipment.

There was good evidence that you could build a good, small racing car and beat cars that were more powerful but not as agile, while having more power than the really small cars had. This theory came together with the Corvair-powered Genie.

When Joe Huffaker designed the Genie sports racer he planned to expand the car, use different engines and sell as many examples as the market could take. The Corvair engine was a natural expansion. Stock displacement was 140 ci and stock power was a rated 80 bhp. The size and weight of the engine and the car worked out, not by accident, to give the compact Corvair the same performance as the bigger Chevrolet sedan with standard six.

The engine for the Genie-Corvair came from Bill Thomas, the same man who did the Corvette-powered Cheetah. By late in 1962 the Corvair engine was a known quantity. Modifications to porting and valvetrain and the addition of carburetors and tuned exhaust could give as much as 200 bhp in race trim, and it was easy to enlarge the engine and still be within the 180 ci, three-liter limit for the SCCA's class C Modified. In this car, stock bore and stroke were kept, and the engine was souped up only to produce a claimed 175 bhp.

The Corvair engine was larger outside than the BMC engine but didn't weigh any more, and of course they could leave off the water radiator. When *Sports Car Graphic* (which by 1963 had replaced *Road & Track* as the boy racer's bible) weighed the Genie-Corvair at trackside, the scales said 830 pounds.

Direct comparison is difficult because the Genie-BMC was weighed dry, in the shop, and the Genie-Corvair had oil and some fuel. But the weights were close enough to believe the claim that the Corvair engine and Porsche transaxle, chosen because it was stronger than the Volkswagen's similar equipment, weighed the same as the smaller drivetrain, and that adding a few tubes and braces hadn't increased the car's bulk.

Installation was easy, the builders said, because the car was already wide, while the flat engine didn't need extra height and of course three cylinders in line on each side are shorter than four.

Sports Car Graphic's track test turned out to be more of a shakedown session. Staff writer (and later editor) Jerry Titus drove the Genie-Corvair, which was sponsored in large degree by EMPI.

Yes, there was an opportunity for conflict of interest.

But Titus was an honest man and reported more than even a new car's share of problems. Huffaker was aware of the latest suspension theory, that the springs and damping should be as soft, and the wheel travel as long, as was practical. The Genies were softly, longly sprung.

The Genie-Corvair had a spare tire in front, fuel tanks on the sides and the driver edging back toward the middle.
Petersen Publishing Company

Titus learned quickly that the geometry of the wheel travel was wrong; that is, where the wheels were pointed and how they were positioned under braking and cornering, as well as the front-rear weight and height combinations, were in error. The geometry changes caused by adjustment to spring rate and weight transfer "promotes some of the most treacherous traits we've ever encountered," he wrote.

Titus did the smart thing and drove one step at a time, getting the braking done before turning, then accelerating through and out of the turns. This was safe and the car could be controlled, but the times were several seconds slower than would be needed even to qualify for a major race.

In the test, which contains no figures for acceleration or braking, Titus described the car and the hope that the weight-power combination would put the Genie-Corvair in the front row.

And so it did, to a degree. But during the next two seasons the car would qualify well, or could keep up with the leaders for part of a race, but couldn't stay there. Some of the disappointment might have been racing luck. The belt for the cooling fan flipped off and the car overheated. Then the drain plug fell out. Then . . .

Sports Car Graphic faithfully reported the car's lack of progress during 1963 and then the Genie-Corvair disappeared, just as its parent Corvair was to do a few years later.

Jerry Titus and the Genie-Corvair, in a picture that tells its own story. Road & Track

Genie Mark 8

Huffaker and partners may have suspected that the slightly smaller engine and car weren't going to be the best way, because eight months after the Genie-Corvair appeared, Huffaker followed with what must be called the V-8 version.

The main bet was on the domestic engine because the new Genie was supposed to compete for first overall in the USRRC and the USAC pro races. Beyond the engine configuration, Huffaker was open for offers. The new car was designed to accommodate the Buick, Oldsmobile or Pontiac version of GM's alloy V-8; or the Chevrolet in whatever size, iron or aluminum if you could get one; the new Ford Fairlane 289 V-8, which was smaller than the Chevrolet on the outside.

The Ford engine had full factory backing for competition because Ford had renounced the 1957 AMA ban on direct involvement in the sport and had decided on the reverse approach. Ford management was using the small-block V-8 as the basis for Indy cars and world-class GT40 prototypes, for Cobras and Super Cobras as Ford-Coopers were known, and it was working with the English Lola . . . it all paid off with wins against Ferrari *and* Chevrolet. Meanwhile, the little Ford V-8 was gaining back much of the reputation lost when Chevrolet took over the performance and youth markets in 1955.

The Mark 8 Genie was an expanded version of the sports racer that began with 1100 cc of power. Huffaker used the same chromoly steel tubing, in the same sizes: 1 in. and 0.75 in. diameters for the main portions of the space frame. The car had more tubes, and a full belly pan of aluminum that was permanently attached to the frame tubes and thus a stressed, load-carrying frame member. Stressed skin panels covered the cockpit section of the frame, and stressed panels were installed in front, at what used to be called the firewall, and in the rear, between the cockpit and the engine bay.

The car had evolved in that the frame came in three sections: the main, center section with cockpit; the front portion carrying suspension, steering and so on; and the rear section with engine, drivetrain and rear suspension. The sections were separate and could be unbolted. In the event of a crash, one section would in theory absorb most of the impact and that

Genie-Olds Mark 8	
Wheelbase	90 in.
Length	na
Tread	51 in. front, 49.5 in. rear
Weight	1,150 lb. trackside
Engine	aluminum Oldsmobile V-8
Displacement	239 ci
Claimed power	300 bhp est.
Top speed	na
0–60 mph	na
Quarter-mile	na
60–0 mph	na

section, the front most of the time, could be removed and repaired, or replaced, with an ease impossible when the damage was still attached to the car.

Suspension for the big car was mostly as seen on the smaller cars. Triumph spindles were used in front, a practice also followed by the English special builders because the small Triumph sports cars and sedans were strong and well designed. Morris Minor rack-and-pinion steering was also used in both the Mark 8 and Triumphs for the same reason. The coil-over shocks were canted so their force was fed directly into the frame. The A-arms in front were widely based and the rear links were so widely based that they were more like trailing arms. The rear hub carriers, as seen before, were cast by Huffaker and were quite tall, the better to put the attachment points where they could follow the correct path of travel.

By now Huffaker and his peers understood suspension, a science that until this period had been mostly an art, and a black art at that. The Mark 8 had fully adjustable settings for camber and caster in front and rear, and the shocks could be tuned for just the right damping. There were springs for basic settings and anti-roll bars for tuning to the track, while ride height and roll centers could be varied to suit the rest of the suspension.

The first Mark 8 used what was listed as the Olds version of the B-O-P V-8, chosen because it was Huffaker's view that the Olds heads used a bolt pattern and port configuration that were stronger and better adapted to improvement.

The Olds was bored to 239 ci and got a hot cam with roller tappets, while the heads were given larger valves and port revision by Traco Engineering, an engine-building firm founded by old-time hot rodders Jim Travers and Frank Coons. The exhaust pipes were all the same length, a technique that enhances power at one engine speed while it knocks off torque at other speeds; systems that mate and match, with pipes joining each other at specific lengths, give less power at peak and a broader power band. Compression ratio was 11.4:1, while atop the engine sat only two, rather small, two-barrel Rochester carbs. Claimed power was just a bit less than 300 bhp.

BMC had solved the gearbox problem by making its own, a beefy transaxle that used Corvette gears for the gearbox and a cast-in-the-USA housing for the rest. It had limited slip for the differential, was cast in magnesium and was supposed to take at least 400 bhp. Because the transaxle was designed from the beginning to do this work, it had the input below the differential, letting the engine be lower than the rear axle's centerline, and putting the gearbox in back of the driving wheels, with the engine thus farther back in the chassis. The Olds V-8 got a light flywheel, only 17 pounds, with a two-plate clutch that had a diameter of only 7 in. The engine could sit as low as the frame, so to speak.

The fractionally larger car got larger brakes, with Dunlop magnesium calipers, and BMC's own hubs and wheels at 15 in. to go with the 11.5 in. discs. This was early in 1963, with superwide tires still in the future,

Dan Gurney and Pedro Rodriguez, both in Genie-Fords but driving for different teams. The Genies were good, strong, reliable cars, if not much for innovation. The Mark 8 model was meant for the domestic V-8 and was scaled to suit, but the design and look remained the same as those of the earlier Genies. Road & Track

so the Mark 8 wore tires of nominal 5.5 in. in front, 6.5 in. in back.

The body, like the frame, was simply a slightly larger version of the first Genie sports racer, with the alloy center panels and front and rear doors of color-impregnated fiberglass. It conformed to the 1962 FIA regulations, with full-width windscreen and genuine second seat, but no real attempt was made to meet state codes or even provide weather protection as seen for races like Le Mans. The Mark 8 was a racing car, period.

The car described in *Sports Car Graphic* was the first of six built during what was hoped to be the first production run. It was a customer car, for Dave Ridenour, a good amateur at the time campaigning a Lister-Jaguar. Presumably because it was already sold to a man who hoped to make his reputation and a few dollars, the car was driven by the owner and the builder, while people from the magazine took pictures and notes. We are told that the tests were at better-than-record times for the track, but we don't know how fast the car actually was.

By deduction, we know the Mark 8 was better than the Corvair version. With fuel and driver aboard, weight distribution was fifty-seven percent front, forty-three percent rear. That's more of a bias than we've seen in earlier cars yet the Mark 8 handled well right off the trailer, in contrast to the Genie-Corvair. Huffaker was taking notes and presumably the tire and suspension people were also learning how to make the chassis work once they had all that power and the traction provided by the weight distribution.

Another form of progress was that Huffaker had gone from being the guy in the shop who had the backing of his company—said company being willing to get some ink from backing the car—to being a producer of racing cars for sale. There was enough demand for parts and equipment to justify expenses like the transaxle and the special wheels and hubs. Genies were, by our definition, graduates from the specials class.

The Mark 8 was a good car. Ridenour was a good driver. The combination appeared in the pro races for the next several seasons, contesting the lead and qualifying within reach of the best. Good but not great, and that praise is not meant to condemn. It also serves who come in third.

The faith defended

While the racing world was clearly moving in the direction pioneered by Grand Prix cars, there were still those who weren't convinced, or perhaps believed that good work would overcome other handicaps.

Dave MacDonald was a talented new Corvette driver who earned his reputation as the bravest of those with "fully prepared" production class Corvettes. When he made his move from production into the modified class, he did it with a sort-of Ol' Yaller.

It's a production Corvette, right? Wrong. It's a Corvette-copy body on a Balchowsky frame and suspension, with Chevy V-8 and Dave McDonald at the wheel. Dave Friedman

McDonald never won much with his sort-of Corvette Special but he did, as shown here, put on a show of bravery and power. Dave Friedman

The car began with the frame from Ol' Yaller IV, which itself began life a little bit bigger than the earliest Ol' Yallers and the subsequent semi-produced version described earlier. The frame, with independent front suspension and live rear axle and drum brakes, got a souped-up 327 Chevy V-8 with a four-speed Corvette gearbox. The body was a modified copy of the body from a 1961 Corvette—the pre-Sting Ray, an overly decorated, four-lights-and-fake-scoops-style Corvette. Why they did it, nobody ever explained.

But, on a 92 in. wheelbase, the car weighed only 1,750 pounds and had about 350 bhp to 400 bhp, and MacDonald was good. The car didn't win much, but it did attract attention and it did earn MacDonald a place in the pro ranks.

Aguila

Interesting for another reason was the Aguila (Spanish for Eagle). This was one of Frank Kurtis' last racing car designs. Kurtis had been in racing since before World War II, with midgets and sprints and Indy cars and sports cars. He was an inventive man,

The Aguila was a Frank Kurtis design for a dual-purpose racing car. Here it has fenders and lights, for sports car racing . . .

always willing to experiment with such things as rear engines and independent suspension, but equally willing to not use such ideas when they didn't work—and they didn't in the forties and fifties. Kurtis had a long run as the King of Indy, then eased off to do work for the government, inside a secret shop fittingly known as the Skunk Works, while his son built boats in the old Kurtis plant in Glendale, California.

But Kurtis was still interested in design, and he did a dual-purpose car for Texas businessman Herb Stelter. Stelter had motorcycle and formula car and sports car experience and he was watching the Indy 500 when he thought, if those cars were so smooth through the turns, they should be good on road courses, so he persuaded Kurtis to design and build a convertible racing car.

Roger Penske did it one way, by taking a Grand Prix car and rebodying it into a sports racer. Kurtis' Aguila went beyond that. It was an Indy design, with ladder frame, live axles and torsion bars front and rear, with Halibrand quick-change differential and cast wheels and disc brakes, but with a wider chassis—shades of the 500K—and with detachable fenders. This was Kurtis' idea, the theory (and hope) being that the car could run in sports car races and the new Formula 366, professional racing for single-seaters with stock-block engines up to 366 ci, or six liters, of displacement.

Kurtis did the frame and body, which worked out to be a sleeker 500 style envelope, similar to the envelope body he'd done for his earlier cars and similar to the last front-engine Sadler. *Contemporary* is probably the best word here. There were housings for headlights in the fenders and the windshield was full width, so the FIA rules were met. The driver sat well on the left, to the rear of the chassis, and the engine, a 327 Chevy, was offset to the right. (One does wonder which of these is the better: the car biased to the inside, or evened out.) The body panels for single- or dual-seat configuration were unplugged by the turn of a Dzus fastener so the car could be converted much faster than the race schedules warranted.

That was one drawback, as the Formula 366 never took hold. There was already as much racing as the public needed or could keep straight, with USAC in formula cars, NASCAR in stocks and SCCA in sports, never mind a mix of the three.

And the new owner wasn't a top driver, judging by the accounts of the races the car won. The Aguila's claimed weight was 1,650 pounds wet, with a shortish 88 in. wheelbase and a wide tread of 56.5 in. front, 57 in. rear. The engine was professionally built and should have had at least 300 bhp, or as much power as the next motor on the grid. But the magazine accounts show the car running not very well and having problems, a new and different one weekend after weekend, while the owner says he'll fix it yet . . . until the car fades away. An old story by now, and it wouldn't be the last time such a car had such a history. As with Stroppe's car or the semi-Ol' Yaller, a special of that configuration and weight and power was plainly out of date, even with a top driver.

Or, wait . . .

The 1963 season displayed the power of progress and the tenacity of good design, backed up by really expert preparation. The battlers in the SCCA nationals and the USRRCs were Jim Hall in the Chaparral 1 (at least in the early part of the season), Walt Hansgen in Briggs Cunningham's Cooper-Buick, Augie Pabst in another homemade Cooper-Buick with Oldsmobile heads, Roger Penske driving the Zerex Special for John Mecom, Dan Gurney in whatever he could get (Coopers and Lotuses with Climax or Chevrolet and eventually small-block Ford V-8s), Chuck Daigh in a Lotus-Climax.

They put on some good races, as the Lotus 19 had become a truly production racing car, with seven such examples at the Mosport pro race for instance. Chuck

. . . And here it's been stripped of fenders and lights and such to be raced as a formula or open-wheel car. Sadly, the car didn't do anything in either form. Road & Track

The configuration was becoming standard, but you could still tell the cars apart without a program. Number 98 is Dave Ridenour in his B-O-P Genie, and number 91 is A. J. Foyt in the Scarab-Chevrolet, at Laguna Seca in 1963. Road & Track

Daigh won that one, 0.3 seconds faster than Jim Hall in his front-engine Chaparral. Small cars (Lotuses and Porsches) outlasted the big ones at Laguna Seca. Ol' Yaller III won the national at Riverside (the pros were busy elsewhere by that time; SCCA's amateur classes had become really amateur again, and yes, it did take some of the fun out of it). A Porsche won the USRRC at Watkins Glen in 1963, with another Porsche second and Ken Miles in a Cobra third and first in the over-two-liter class, for a switch on the usual order.

But the Genies were working. Pedro Rodriguez won at Kent in a Chevy-powered Genie Mark 8, and the Genie-Olds built as the prototype was third, driven by owner Dave Ridenour. This was Huffaker's (and backer Kjell Qvale's) first main-event win in three seasons and it was welcome. There was no question that the smaller cars could do well on the tighter courses but even so the sheer power of the Chevrolet and the power-to-weight ratio of the B-O-P and Ford engines would prevail. Ford, the corporation here, was fully backing all forms of racing and through Carroll Shelby ordered a fleet of Coopers with Ford V-8s, to be known as King Cobras.

As Chairman Mao said about something else, there were 1,000 configurations contending. We are building toward the peak, of American sports racing, but before we arrive we can survey what was being done, and why.

Begra

Gene Beach and Henry Grady, whose names easily combined to become Begra, were top builders whose struggles and results typified their times.

They lived in Florida, about 300 miles apart, which made working together something of a commute. But they worked well together and had com-

At first the Begra was a bit rough at the edges, and used Fiat parts and power. That's partner Henry Grady at the wheel.
Road & Track

Later the Begra used Saab power, with the front-drive engine mounted in the rear. This later body had more provision for rules, as in the covers for higher headlights.
Road & Track

plementary skills and resources, so it was worth the effort.

Their special projects began with the purchase of a new Fiat 600 sedan, which they took to the shop, tipped on its side and stripped. They put engine and drivetrain aside. Then they attached steel plates to the shell, and put conduit on the plate until they had a jig and could build a space frame with all the various suspension and mounting points where they had been (using the factory's geometry seemed better than guessing). Then they removed their additions and sold the body shell back to the dealership.

The new frame got an aluminum body, the engine was bored and stroked from 600 cc to 750 cc, for class H, and the new car won its class right off. Then the team hit the problems most special builders had. The Fiat engine wasn't as powerful as the racing engines were.

The first car was sold and a second built, this time with Saab engine. At the time Saab was famous for its hot rally cars and for building the best two-stroke triples in the automobile world. But the ordinary buyer had trouble getting the hot engine, the GT version, so Beach and Grady took a sedan engine and borrowed the GT version and copied it.

Or tried to. Two-strokes are not like four-strokes. The modifications consist of changing the size and shape and location of the various ports that let mixture into the crankcase, then from the crankcase to the cylinder and then to the exhaust pipe. So Beach and Grady had to learn how to make a new type of engine work, and they had to adapt the front-drive Saab engine, which in stock location projected in front of the front wheels, to the Fiat transmission, which was for a rear-drive car and which had its original engine sticking out aft of the rear wheels. Grady owned a parts store and used to say their work was possible only because he had all those shelves in back. He could take calipers and micrometer and run into the stockroom and measure bits and pieces until he learned, for instance, that you could mate a Ford engine with a Fiat gearbox by using Austin clutch plates.

Late in the Saab-Begra project they did get a genuine GT engine, except that it arrived just barely in time to be fitted to a new, stronger frame and run at Sebring in 1960. The car would barely run and the Saab representative who was supposed to help got there after the race began. The car limped around for a few hours and retired.

The team built a third car, with a twin-cam Abarth-modified Fiat engine, which was much more powerful, and Beach was able to score class wins. Meanwhile the second car was converted to Ford power.

Ford of England was similar to Ford of America in that it produced ordinary sedan engines that were converted for racing, then looked at racing as a means of promotion and as tests of engineering. The little 105E four had an ultrashort stroke and generous bore, and a unique, hollow and strong crankshaft. It came with overhead valves and became the tuner's pet. Next were larger versions, 1100 cc, 1300 cc and 1500 cc. Then Colin Chapman did a twin-cam conversion for the 1500 Ford and put that into the Elan and the racing cars . . . until Ford backed a project that took two of the four-valve, twin-cam engines and joined them at 90 degrees to make the Cosworth DFV (double-four-valve) V-8 that dominated Formula One and later, turbocharged, ruled Indianapolis.

True, there wasn't a whole lot of the original overhead-valve four in the engine when it reached final form, but the line of descent was clear and direct. The English Ford deserves to stand next to the Ford flathead V-8 and the Chevrolet Mouse Motor in the Hall of Great Racing Engines.

Armed with the Ford engine, the Begra was a good example of what became the club racer, built one at a time for amateurs and owners who simply wanted a competitive racing car. Beach himself would go on to produce a line of competitive formula cars.

Jabro Mark III

James Broadwell followed a similiar path. He began with a homemade road-sports car, then did a fiberglass body atop what was essentially a Crosley engine and running gear. Rather than building complete cars, Broadwell's business evolved into supplying kits. Jabro produced the parts: a space frame, a fiberglass body, seats and various bits of hardware that would be difficult for the homebuilder to find locally or make. At the other end, it was less expensive for buyer and seller if the hand labor, especially the easy parts like bolting stuff together, was done at home.

By late 1961 Jabros had gone from the basic stage through better streamlining and suspension to a clean and sophisticated design, saving only that Broadwell was still using the Crosley as his basis for an engine, and that the engine was still in front.

The Mark III was the final such from Jabro. It was a tiny, basic car, with live rear axle and independent front suspension; it used the old Allard idea of a beam axle cut in half and pivoted at the center, a swing axle for the front wheels. This was simple to make and because the Crosley steering could have its linkage in the center, aligned with the axle pivot, suspension geometry could be kept in track, so to speak, with wheel travel in an accurate path. The car had left-hand steering and the engine was offset two inches to the right, for balance and for extra legroom.

By 1961 the Crosley engine was almost totally replaced, part by part, with racing gear. Jabro offered an eight-port cross-flow head so the engine could take advantage of the revs permitted by the overhead camshaft and the ability to ram tune, with exhaust and intake lengths chosen to extract extra power at a given engine speed. Thus, the claim of 65 bhp, nearly one and a half horses per cubic inch only four years after Chevrolet's boast of one for one.

Beyond that, the Jabro kit would work with any of the other engines in use then—the Saab or Fiat or even Mercury outboard—or, as time passed, with the English Ford. Turning things around, one buyer kept most of the car but juggled and put the engine in the rear, with his own independent suspension and transaxle.

Jabro did a good business in the kits, although by this time the front engine and the suspension, at either end, were out of date and the model wasn't capable of a national win except perhaps in exceptional hands.

Dolphin-Porsche

By 1963 road racing had become one part of a worldwide sport, with crossovers in several directions, often at once. We saw some waxing and waning, as the specials developed into, well, specialized vehicles for American rules and courses, then were set back by new ideas from elsewhere, with the lightweight Lotus-Elva-Porsche contingent taking the small classes and sometimes overall.

The Dolphin-Porsche was a counterpunch.

Dolphin was a California builder of single-seaters, one of many that got into the business when Formula Junior, which required small stock-block engines but allowed anything else, was at its peak. When the peak was over, Dolphin's owner wanted to get into modified sports classes.

Otto Zipper was a Porsche and Ferrari dealer and a racing fan. His driver in 1963 was none other than Ken Miles, whose presence at the wheel of racing Porsches showed that the Porsche was the car to have and that not even Miles was ready to tackle the factories with what could be done in the backyard.

The racing Porsches were built for 1,000 mile races, however, and were heavier than American short sprints called for. So, with Miles to oversee and Dolphin's designer to draw, Zipper commissioned a sports racer with single-seat components and a 1700 cc racing Porsche engine, the lot covered by a body that met the FIA's 1962 code for Group C, or sports cars.

The frame was a triangulated cage, or space type, of 0.75 in. steel tubing, with a stressed aluminum belly pan. The builders said the complete frame weighed 100 pounds, which was heavy by comparison with other space frames. The second claim was a deflection stiffness of 3,000 pounds per degree: it would take 3,000 pounds of force to twist the frame one degree

Jabro Mark III	
Wheelbase	83 in.
Length	na
Tread	46 in. front, 44 in. rear
Weight	750 lb. trackside, est.
Engine	Crosley
Displacement	45 ci
Claimed power	65 bhp
Top speed	na
0–60 mph	na
Quarter-mile	na
60–0 mph	na

Dolphin-Porsche	
Wheelbase	89 in.
Length	na
Tread	51 in. front, 51 in. rear
Weight	1,100 lb. trackside
Engine	Porsche RS
Displacement	1700 cc
Claimed power	165 bhp
Top speed	na
0–60 mph	na
Quarter-mile	na
60–0 mph	na

out of alignment. (When such claims became more common, a selling point in some instances, other designers pointed out that any figure so used was a calculated guess. Nobody was actually twisting frames and measuring their resistance because that would damage the frames. Still, it's an interesting figure.)

Steering was good old Triumph Herald; front suspension was fabricated upper and lower A-arms. All the connections were aircraft-type Heim balljoints, more than forty of them, rather than bushings; the idea was to eliminate flex and thus completely control wheel travel and location.

Rear suspension adapted the Dolphin magnesium hub carriers to the RS Porsche transaxle. With extra power, the hub carrier needed to be wider by 2 in. A wishbone (A-arm) was at the bottom and a single link was at the top, with trailing arms top and bottom. The Armstrong coil shocks could be mounted at different angles, which effectively changed spring and damping rates, and there were adjustable anti-roll bars at each end, while the front and rear roll centers could be moved up and down in 0.75 in. increments. The art of suspension tuning had come a long way since Kurtis and the Indy crowd taught the sports car set how it was done.

Dolphin cast its own wheels, 15 in. ones in this case, with 5.5 in. rims in front and 6.75 in. rims in back. There was a constant compromise here, as by now the benefits of wide rims were known but they required extra care in selecting the path and thus the effect of suspension geometry. The taller wheels and tires gave extra space for bigger brake discs, 11 in. ones in this case, but made the car higher and increased frontal area and thus slowed the car. As cars became more complex, so did the decisions needed to make them work right.

The Dolphin-Porsche's engine was no problem. By 1963 Porsche was *the* builder of semi-production engines in the middle sizes. The RS engine was an opposed four, with double overhead (OK, outboard) camshafts on each bank. It was air-cooled with a big fan atop the crankcase, and because the configuration was short and wide it fit the space perfectly, while not intruding anyplace. And because the engine was designed for a rear-drive chassis, Miles and Zipper could simply go with the excellent transaxle and most of the drivetrain that came with the RS.

The team took care to make the fiberglass body and the configuration meet Group C rules, with required fuel tank placement and doors and windshield, while if they so chose they could have fitted the

The Dolphin-Porsche was a logical approach in the timeless theme of a good engine in the lightest possible chassis. Style was seemingly based on the early pontoon-fender Ferrari Testa Rossa. Petersen Publishing Company

car with a larger windshield and added some ballast and run the car as a prototype, in the world championship class. In SCCA form the Dolphin-Porsche was about 150 pounds lighter, and lower, so all the time and money were well spent.

Or perhaps there was a jinx involved with telling the magazines all about the car before it had wiped up the competition. Driven by Ken Miles and Lew Spencer, who was a good man from the production car ranks, the Dolphin-Porsche did well, but in the long run lost out to cars that were not as close to the FIA and tuned more fiercely, or something.

Harrison Lotus-Ford

It may have been that by this state of the special's art, engine swaps were no longer enough.

The Lotus 19 was an excellent sports racer and Colin Chapman knew how to gauge the market and

The twin-cam air-cooled boxer-four Porsche engine was an easy fit in the Dolphin's space frame. Petersen Publishing Company

Revisions to the frame and body as well as the new engine made the Harrison Lotus-Ford, shown here with Indy star Lloyd Ruby, more than another engine swap. Dave Friedman

meet demand, so the grids were full of Lotus 19s when the professional series was born. They came from the the factory with a choice of Coventry Climax fours, 1.5 to 2.7 liters most of the time, and they were good little cars.

When tires and suspension and availability of things like transaxles and disc brakes put racers into the stage where a good big car could always beat a good little car, the good little car wasn't enough. So J. Frank Harrison, a well-to-do enthusiast who drove his own cars in regional SCCA races and sponsored professionals in the pro races, used his Lotus 19 as the basis for a better, bigger car.

Harrison benefited from Ford's corporate program. The engine for the Harrison Special was the small-block V-8, as first seen in the Fairlane sedan and then developed with the Cobra and with Ford's Indy program. The 289 in the Harrison car used a factory racing camshaft, the normal reworking of heads and internals by Traco, and Weber carburetors as fitted to the Cobras, with an exhaust system developed for the Indy cars—free gifts from the viewpoint of those who had to develop this engine.

The frame was Lotus, with the engine bay reworked and with various gussets and triangulation where indicated. Part of that may have been insurance. On the one hand, the Ford engine wasn't any heavier than the 2.8 Climax it replaced. On the other hand, Lotus was famous for not using any more material, and thus not making the car any stronger, than it absolutely had to. Tuner Jerry Eisert surely knew this and beefed up the space frame accordingly.

The engine was of course behind the driver, and the transaxle was behind the engine. Eisert used a four-speed Colotti—the most expensive of the several brands, such as BMCD and Hewland, then on the market, but one that had features like pressurized lubrication.

Because the drivetrain was different, the independent rear suspension had to be revised. The driveshafts from the new gearbox were shorter than the old ones, and served as the upper control arms, so the geometry of the rear wheels' travel was different. That should have meant the wheels' camber (the tilt they take when they move under compression) was different. And it must have been, but it didn't affect the handling any and the rest of the system—the lower link and the hubs—was as it came from Lotus.

Brakes were disc, by now standard on every special of any consequence, 12 in. in diameter in front, 11 in. in rear, with 15 in. cast alloy wheels. The driver was on the right, and the engine was centered.

The new engine was more powerful and needed a larger radiator, a cross-flow aluminum one with core from a racing Corvette.

Troutman and Barnes made a new aluminum body, with a narrow aggressive nose separate from the front fenders and a lot like the Ferrari Testa Rossa.

None of the magazines clocked the car or took it apart, but Eisert's estimates were 325 bhp for the engine, a few pounds less than 1,400 as racing weight.

That should have put the Harrison into contention, and it did. One more advantage was Lloyd Ruby, who was a good USAC driver and the USAC stalwart who became the best of the converts to road racing.

Ruby and the Harrison won the 1963 Kent professional race, in a walk and against the top teams.

Using the Ferrari V-12 instead of a domestic V-8 made the Team Rosebud Lotus a different mixture. On paper, the combo worked and driver Innes Ireland, shown here at the Times Grand Prix, 1964, was world rated. But an early accident put the team behind the times and the car never lived up to its promise. Dave Friedman

The car was up to the competition, even though it took more than a simple engine swap to make it so.

Rosebud Lotus-Ferrari

This was another example that proved the rule, and one that shattered a cliché.

Yes, one would grin ear-to-ear with such a nifty little creation. Slick Mercury Outboard Special doesn't have the driver in the middle, either; it's just that Dan Odenburg fills the cockpit. Road & Track

Rosebud Racing was a Texas-based sports car team, which also owned a Lotus 19, and which also came to realize that the Climax four wasn't up to the job any more.

If a V-8 from a sedan is good, then a V-12 from a racing car—an imported racing car—had to be better, right?

Not always. The Rosebud crew followed the standard pattern. The 10 in. discs were replaced with 12 in. discs, the frame was stiffened and the engine mounts were revised and improved. The more powerful car got wider cast alloy wheels and stronger uprights for the rear suspension and a Colotti gearbox mated to the Ferrari engine, in itself a major machine shop job because the engine was originally fitted in the front of the parent car.

The Ferrari engine displaced three liters and was rated at more than 300 bhp, with no exact figure given—or to be relied upon, as the Italian makers were famous for finding more horses than did other people. The V-12 was larger than the V-8, and because it was equal power from more engine speed and less torque, the Rosebud Lotus used a five-speed gearbox rather than the four-speed preferred in domestically powered conversions.

The Ferrari engine was heavier than the Ford, but again we have no independent figure. Steering and driver remained on the right, with centered drivetrain and with saddle fuel tanks, one on each side of the centralized cockpit. (By now the FIA's rules were so cheerfully accepted, and aerodynamics so well-known and appreciated, that few builders even bothered with little tricks like passenger seats that couldn't be occupied. It simply wasn't worth the effort.)

The Rosebud Lotus kept the stock body, with some minor changes. The team hired one of the better guns, Grand Prix driver Innes Ireland. The combination was only a fraction off the pace for the last three USRRC races in 1963, but it didn't win. At the third event, Kent, Ireland drifted into the loose gravel and debris that gets swept onto the edges of racetracks. He lost the car and careened into a course worker's car parked in the escape road. Ireland wasn't hurt but the car was badly bent. It was repaired but somehow never showed as much speed again.

On paper, the Lotus-Ferrari made sense. In fact, it was an interesting combination that didn't click.

Mercury Special

This little charmer probably should be viewed as a fascinating project instead of an object lesson or a way to go racing, because it didn't meet the FIA rules and it never won any major events. Even so . . .

The Mercury Special began with a different Mercury: an outboard motor, an inline four-cylinder of the two-stroke persuasion. Outboard motors and two-strokes go well together. They were then (and are now) a secret; most outsiders don't know that the

Mercury outboard engine went in the right rear corner, lying down across the chassis. The four expansion chambers are critical to proper tuning of such an engine, and the checkered flag in the cockpit implies that the car has just won again. Road & Track

boat racers, or even the production engine buyers, are using V-6s and V-8s that crank out twice the power per cubic inch of anything you can buy on four wheels. They are as exotic as the world-class motorcycles, which have V-4s with 4 bhp per cube, while the outboards are sold in stores and the motorcycles come only to those on the factory team.

The debit side is that two-strokes live in their own world with their own rules, and they don't always respond to rational thought.

Two Oregon builders, Dan Odenberg and Scott Hamilton, began work in 1960, discarded their first attempt, thought about it and started over. Their first completed car, raced in 1963, was tiny, with a wheelbase of 64 in. and a tread of 43 in. front, 46 in. rear. With fuel and oil and so on the finished car weighed 490 pounds on the grid, and with the H class engine producing probably 60 bhp the power-to-weight ratio was impressive.

The builders made their own space frame of steel tubing with stressed alloy pan and fiberglass body, in a clean wedge shape. The fabricated A-arms for the front suspension and trailing arms with coil-over shocks for the rear, with the driveshafts serving to also locate the rear wheel.

The Mercury Special could be so small because the engine, an inline four, was crossways in the rear, with a BSA motorcycle gearbox and clutch tucked next to the engine and driving a chain to the rear sprocket that attached to the driveshafts. The builders made use of various bits from a BMW Isetta mini-car, such as 7 in. drum brakes and some drivetrain parts. A sign of the times, perhaps, was that where at one stage the builder of a homemade racing car couldn't get parts big enough, this car needed parts smaller than those the other cars used.

This was a nicely constructed car and the team did well with thinking out the mechanical components. There was a problem, though, in that the rules changed between the time they began and the time they had the car actually running. The car wasn't legal for the FIA rules that came into effect in 1964, so it was raced locally and sold as a club racer: in those times the local chapter of the SCCA sometimes could help overlook minor contortion of rules imposed from elsewhere.

Bobsy Mark II Ford

The Bobsy story is a success story, as well as an object lesson in doing what works best.

Bobsy was Jerry Mong, an Ohio enthusiast who built two sports racers with the help of friends who wanted to go racing. They used German DKW engines from the mini-car by that maker. Although the cars were reliable and handled well, they weren't fast enough to win.

Mong decided he liked building racing cars well enough to do it as a business, so he obtained financial backing and designed and built what became a prototype. It was conventional for the day, with engine in the rear, space frame and so forth, but it was powered by an 1100 cc version of the English Ford.

Mong had a top driver. Chuck Dietrich was an experienced driver of small modifieds, and he helped shake the car down. In 1963 he drove the Bobsy to five class wins, two seconds and one third, and the Bobsy became the first American-made car to win the SCCA's G Modified class. With that title secure, the team swapped the 1100 cc Ford for a 1500 cc version and raced the same car in F Modified, with more class wins. Dietrich used the bigger engine to place fifth overall in the Road America 500 in 1963, and was second in class at the Mid-Ohio USRRC until the car lost cooling water.

Convincing stuff, enough to put Mong into the car construction business with production versions of the vehicle.

He began with aluminum tubing for the space-style frame. The tubes were about the same size as they were in steel frames, with slightly thicker walls, and with fabricated alloy sheet bulkhead and stressed panels for extra strength. Mong told *Car and Driver* that given skill and heliarc welding, and the proper planning, a car was as easily framed in aluminum as steel, so much so that he didn't understand why the other folks didn't do it his way. The frame, Mong said, weighed only 44 pounds complete.

That material aside, Mong didn't do much experimenting. Front suspension was upper and lower A-arms, with coil spring-shocks and with rack-and-pinion steering. In the rear it was a single arm for the upper link and a wide-based A-arm below, with cast uprights for the hub carriers and the shock-springs angled to allow for body roll. Again, all the settings for camber and caster and roll center and ride height and stiffness of the front and rear anti-roll bars were adjustable. By 1963 nobody would have bothered to make a road racing car without such indispensable tuning aids.

The disc brakes used 9.5 in. rotors. The cast wheels were 13 in. in diameter, 6 in. wide front and back, although the rear tires were wider than the fronts, 5.50 versus 4.50 in.

The engine was off-the-shelf, the shelf being English tuner Holbay. The Ford 105E was a strong little unit with virtually unbreakable crankshaft and cylinder block, with overhead valves and with two ports for every cylinder.

That sounds normal, right? The English didn't always think so. BMC engines, as seen in MG, Austin and company, most often had long strokes and flexible cranks and used siamesed intake ports, shared one and two, and three and four. This saved production bothers. It also meant the engine didn't lend itself to tuning in the racing sense. You can't have one carb throat per cylinder if the cylinders don't have their own ports, and you can't tune with intake and exhaust length unless each cylinder has its distinct path.

Scores of good tuners worked with the MG and Sprite engines, and even the little sedan units, and they did get power and win races. But after years of toil over the siamesed BMC engine, to obtain a Ford and strip it and see that sturdy little crank, the generous bore, the lovingly shaped ports... oh dear. Call it an answered prayer.

There arose a cottage industry of firms making every possible improvement for the 105 and larger siblings. The list is too long to reproduce here. Call it one of everything.

Mong's Bobsy used an 1100 cc Holbay, with a pair of side-draft Webers and the exhaust pipes sized so the primaries were of a set length, then one and four joined and ran next to two and three (which were also joined) for a certain length, then those two pipes became one and that ran for a certain length. *Extractor* is the term car buffs use because the exhaust pulses reinforce each other and the pulses create a vacuum and pull the exhaust gases out of the cylinder when the valve opens. The same thing happens on the intake when there's a separate pipe for each, so you have 1200 cc of fresh mix flowing into the 1100 cc engine. Wonderful stuff here.

The Holbay-Ford came with 100 bhp, right out of the box. Said engine bolted to a Hewland transmis-

Bobsy Mark II Ford	
Wheelbase	88.5 in.
Length	na
Tread	49 in. front, 49 in. rear
Weight	740 lb. trackside
Engine	Holbay-Ford
Displacement	66 ci
Claimed power	100 bhp
Top speed	na
0-60 mph	na
Quarter-mile	na
60-0 mph	na

sion, another English development based on the everlasting Volkswagen Beetle transaxle. It was strong enough for a year's worth of 100 bhp, and the gears could be juggled to suit the engine's characteristics.

The Bobsy had rubber U-joints rather than metal, to allow for rear wheel travel while giving enough flexibility to make up for misalignment or variation in engine mounting and wheel deflection.

Mong's sense of daring was achieved by the use of aluminum rather than some exotic technique. The frame and bulkheads were reinforced by stressed paneling in the front and sides, presumably allowed because the material was so light for its strength. (Designers wanted the stiffest practical chassis-frame so the suspension's performance could be predicted.

Bobsy team car was craftsmanship personified, with perfectly finished and painted fiberglass body. Road & Track

When you have an immovable platform, you can predict where the suspension will place the wheels in any given loading. The old way, with flexible frame and inflexible springs, couldn't be calculated, so nobody knew what worked.)

Mong's strong chassis allowed him to use thin fiberglass sections for the body, designed to FIA code and with such little niceties as real headlights under streamlined covers and adjustable seat and pedals so drivers of various sizes could be accommodated.

The Mark II was supposed to be produced for sale, with a choice of engines, and Bobsy had a good production run with the model. For the next several years, in fact, the brand was one of the better sports racers.

The inauguration of the USRRC in 1963 settled both the debate over whether the SCCA should have professional racing—with a resounding yes!—and the question of whether the SCCA would prevail as the organizer and sanctioner of road racing—again, yes. There were two counts here, as USAC retreated to the oval and stock car venues, and the SCCA absorbed Cal Club, the original force for sports car racing in southern California. The Los Angeles Region of SCCA became the Cal Club region.

As further developments, SCCA had adopted Cal Club's rules for production racing. The catch name here was Prodified, meaning that production cars could be modified with hot cams and additions of various options, while retaining stock valve sizes, stroke, size and type of carburetor, and a host of other restrictions. The hope was to make it easier to police the racing by having the rules less restrictive. Then, to keep things in balance, the production cars were assigned to classes based on performance.

What does this have to do with specials? Indirectly, a lot. *Car and Driver* ran a nice essay early in 1963 showing that since preparing a production car

Bobsy's frame tubes and bulkheads are aluminum. Engine is Holbay conversion of English Ford, with two Webers carbs and tuned 4-2-1 exhaust. Note bolt-in triangulated brace across right rear of engine bay, and brake light at rear center above transaxle. Road & Track

for racing had become more expensive because of the changes allowed, racing with a real racing car had become less expensive.

(For as long as there's been road and sports car racing there's been a nostalgia for the days when you could drive to the race, tape your lights and remove the spare tire, race and then drive home. In truth there was never a time you could do that and count on going home with a trophy, Max Balchowsky's case excepted.)

Along with the new production rules, the SCCA tried to make things easier for national championships by dividing the United States into two halves, and then proclaimed a "national" champion for each class in each half. The 1963 champs in sports racing classes included Harry Heuer in the Chaparral, Roger Penske in his...um...Cooper and Chuck Dietrich in the Bobsy in the East. In the West a Lotus 19 with Buick V-8 edged Ol' Yaller III in C Modified and a Genie-Climax took G Modified; the others in class on both coasts were Lotus and Porsche and so on.

The awkwardness of having regional national champions inspired the people at *Sports Car Graphic* and they inveigled their owners into allowing the magazine to sponsor (meaning invent and conduct) the first really national championships.

This is one of those great ideas that sounds easy, once somebody else thinks of it. Points were awarded for each class in each of the SCCA's six divisions (i.e., North, Midwest and so on). The drivers with the most points in class were invited to come to Riverside at the end of the year, to race against the other divisional leaders. Yes, it was expensive for the racers from the North and East to haul themselves to California, but *Sports Car Graphic* came up with a towing fund, with money from the qualifying races to be saved up and paid, so much per mile, to the leaders.

It was fair, and it worked, although some quibbled that having the titles depend on one race left too much to chance. The other snag was that at first the SCCA didn't consider the winners of the run-offs to be the national champions, so one person could win the race against the leaders, and a driver that person had beaten could have the title.

To the SCCA's credit, the directors quickly saw how much the drivers liked the run-offs, and how well they worked and how fairly the system came out, so the title was awarded to the run-off winner.

McKee Chevette

Another long-term result of the run-offs and the new way to award national heroes was to separate the top amateurs from the pros. The divisional races were fair and close, but for the fan who wanted to see those he or she had heard of or read about, the pro races were where to spend the discretionary dollar. (That and rising insurance costs, or so the SCCA said, would within a few years revert amateur racing back to a nonspectator sport.)

The impact of the professional series did make the stars household names—well, in some houses anyway. This caused new people to enter.

Sponsorship was becoming viable. When Penske first entered his original special, there was one occasion for which he had to tape over the sponsor's name. But in part because of tradition in American racing and in part because it brought in fresh talent and backing, the notion of using racing to gain attention became accepted.

Thus Dick Doane, who owned a Chevrolet dealership in the suburbs of Chicago, commissioned a special from Bob McKee, who ran a machine and fabrication shop in the Chicago area. McKee had been in stock car racing and had crewed at Indianapolis. He had impressed people there, which is how he got the job of converting a Cooper Monaco to Buick power for Rodger Ward. When he did that, he built his own transaxle based on the Corvette four-speed and Sting

Size isn't everything. At the ARRC 1964, Dick Doane in the McKee-Chevy is about to be passed by Jerry Titus in the Webster Special. Titus will go on to win the D Modified class. Road & Track

McKee's original racing car, built for a Chevy dealer, had so much promise and there was so much demand that there was a short production run of five cars. This is one of the later examples, built for a Dodge dealer with help from Chrysler Corporation, and powered by no less than a 426 Hemi. Alas, the engine was too big for the car or perhaps the course, and it had an undistinguished career. Road & Track

Ray differential with quick-change. That project introduced McKee to Chevrolet people, leading to the commission for Doane, and gave McKee a close look at current sports cars, as in the Cooper.

The arrangement the partners reached was also typical of the day: McKee was to do the design with as many Chevrolet parts as possible and could use all he learned to make cars of his own, for sale, as Jim Hall and Troutman and Barnes had done not long before this. Doane's example was to be known as the Chevette, a mix of the Corvette and the Chevelle, Chevrolet's compact model at the time. (This Chevette is not to be confused with the later Chevette produced by General Motors.)

McKee's design was mostly inspired by the Cooper. It had a space frame of steel tubing, round in most places, rectangular when forming the front and rear bulkheads. The front of the driver's compartment was framed by magnesium panels, with the front panel serving as protection for the occupant and a mounting place for the controls. With brackets and engine mounts in place, the frame weighed 117 pounds.

The front and rear bulkheads also served as mounting points for the suspension. There were series of mounting holes, so the roll centers could be varied. The rear suspension uprights and trailing arms were modified Corvette. The front suspension, independent with upper and lower A-arms, was part Buick and part Chevelle. Steering was rack and pinion, from a Sprite. (Can't have everything, eh?)

The fuel tanks, one on each side for 48 gallons total, were outboard of the cockpit and were rolled from aluminum as part of the body. The body itself, also aluminum, was normal for the time except that the front center was open above and aft of the radiator for an extractor effect. *Sports Car Graphic* described the McKee Chevette as a sort of Harrison Special, sort of King Cobra; that is, Cooper-Ford. The driver was on the left, in this example, with the engine and drivetrain centered.

By 1964 racing parts were easy to find. The McKee used 11.5 in. Girling disc brakes on all corners, with cast-magnesium wheels, 15x7 in. front, 15x8.5 in. rear. The wheels came from American Racing Equipment, which had gone into business with magnesium, and later aluminum, cast wheels for sports as well as racing cars and even the new domestic performance cars.

Parallel to this, Ford and Chevrolet were competing on the track. The street version of the Corvette was semi-secretly supplemented by the racing version, the five lightweight Grand Sport cars from the former racing shops and sold only to racing teams and sponsors for people like Hall and Penske, and like Doane, who had one of the rare coupes early on. (The Grand Sports were almost but not quite a match for the Ford-subsidized Cobras, which in turn filled in as sports cars after Ford's Thunderbird grew into a luxury land yacht.)

Doane supplied the engine—a Chevrolet, naturally—a V-8 with cast-iron small-block. It was bored and stroked from 327 ci to 363 ci, as big as they dared go then, and used tuned exhaust and a modified version of the fuel injection seen on the Grand Sport engines. *Sports Car Graphic* said it was told the engine had produced as much as 460 bhp on the dynamometer, which sounds high but not by much. McKee's own transaxle was used; this too was for sale to other builders, so BMCD, Hewland and Colotti had another rival.

The drive from engine to gearbox went under the differential and then forward, with the gearbox itself placed aft of the driving wheels and the engine's crankshaft allowed to be 2 in. lower than the centerline of the wheels. Another development forced on tuners was that the lower engines and the increased cornering speeds of the cars centrifuged oil away from the intakes to pumps with the production oil pan below the engine. This caused the builders to convert to what's called dry-sump lubrication, with the oil kept in a tank separate from the engine and located where the weight will help the balance. Two pumps are used, one to feed oil from tank to engine and the other to scavenge oil from the bottom of the engine and return it to the tank; hence the term dry sump.

The McKee-Chevrolet and McKee the builder went on to honorable careers in the USRRC. They are presented here as an exhibit of what most people were doing; the Mckee cars were state of the art in 1963.

Chapter 7

The Special of Specials: Chaparral 2

Jim Hall's fantastic plastic: The best-laid plans become racing's best-kept secret

For a long time, almost every special began with the same idea, that the best place to hang the engine and suspension and body, and the best way to keep the wheels from hitting each other, was to build a framework of tubes, a space frame, and attach everything to it.

In 1962 Colin Chapman stunned the racing world with a tubeless frame. The Lotus 25 was a Grand Prix car, and the part that served to hold the engine and keep the wheels in their respective corners was stressed skin. The formal name for this is monocoque and the formal definition isn't quite met in racing. A true monocoque, virtually the only true monocoque, is an egg. An egg has a stressed skin all the way around, enclosing what it holds, and all the structure's strength comes from that stressed, rigid skin.

The next step after the total monocoque would be the full-coverage helmet, as seen with Bell's Star. You have an opening for the head to go in, and a window so you can see. Otherwise, the head is enclosed, and the helmet that goes all the way around is stronger than one that's simply a cap.

This principle easily applies to cars. Well, not easily in practice. BRM used stressed skin, in places, with Grand Prix cars as early as 1955. Phil Vincent's postwar motorcycles hung the headstock and front suspension off the oil tank and the engine below the tank and the rear suspension off the back of the engine, which made a frameless motorcycle. And we've seen a score of specials with stressed bulkheads and firewalls and belly pans, so using the exterior as the skeleton wasn't something Chapman pulled from thin air. Nor was the Lotus 25 a total monocoque. They called it the tub.

That's what Chapman invented. Other builders had used belly pans and saddle tanks and firewalls. Chapman took the metal sheets and wrapped them and riveted them into one structure, the lower half of the racing car with tanks and firewalls and bulkheads and so forth all one piece, the skin serving as the frame. The suspension bolted to the front, the engine to the back, the gearbox to the engine, and the rear suspension to the engine and gearbox. The top half was the body proper, the aerodynamic part.

Chapman's daring and brillance paid off, again, and the monocoque tub became how the Formula One and Indy builders did it. Not all at once, however, and other things were going on.

The next block in the pyramid was the networking among the better builders and drivers, call them

It's the start of a new game, one could say, as Jim Hall and the Chaparral 2 smoke off a field of Coopers and Genies at Riverside, 1963. Petersen Publishing Company

Quietly confident, Hall strikes a Texas pose at the wheel of his radical new car. Petersen Publishing Company

The Chaparral was designed with a plain front. However, early tests revealed that the car was floating on high-speed air, so Hall's team designed an air dam, which by this stage had had an addition tacked onto its lower edge. Petersen Publishing Company

Roger Penske and Jim Hall, and the quiet men in Chevrolet research. Penske was a sales engineer for Alcoa and served as the conduit for technical data and parts when Chevrolet needed aluminum versions of the Mouse Motor's block and heads. Chevrolet and Corvette people were building experimental cars, such as the CERV-1 rear-engine single-seater and the Grand Sport Corvettes. They were banned from direct involvement in racing by the company, but they were allowed to talk and trade parts and arrange for other people to buy and race the products of their experiments.

This was delicate work. They were walking lines so thin as to be nearly invisible. At the same time the motoring press was collectively as much racing nut as crusading journalist. Those in the business knew more than they put in print, and pretended to know more than they had been able to learn. Nor has all the mud in those waters settled to the bottom after twenty-five years.

With that in mind, we return to Jim Hall and partner Hap Sharp.

When Chaparral Cars commissioned Troutman and Barnes to build that first run of front-engine sports cars, they knew the design was just about to become obsolete. They went the old route because they didn't have the best parts, namely a transaxle to put the engine aft of the driver. The Chaparral 1 won because it was light and powerful and fast and reliable and well driven, and because the other teams weren't perfect, either. But even during the Chaparral 1's time at the top, Hall was thinking and working out the next stage.

Another consideration was material. Harking back in time still another step, there was the belief that fiberglass was the material of the future. The Corvette and specials like the Devin, though, were the only manifestations that got beyond the show car stage.

Hall reckoned to change that. He knew an aircraft engineer named Andy Green who'd gone into the boat business, using fiberglass as the material for hulls. Hall had his engineering degree and several seasons of racing and his empirical experience with the Chapar-

You can see the suspension but you can't see the frame because there isn't one. Instead, the suspension and radiator were attached to a fiberglass-reinforced plastic box, which was glued to a larger box holding the driver. This was the first appearance of the first Chaparral 2. Later the coolant pipes and such were routed below the footwell and cockpit. Petersen Publishing Company

ral 1, so he and Green sat down and designed, then built, the first fiberglass-hulled sports racer ever.

Why? Hall said first, it was less expensive and easier to use fiberglass (the correct name is FRP, for fiberglass reinforced plastic) than metal. The plastic could be laid into a mold and reproduced, it was strong for its weight, it had no fatigue life that had yet been measured and it wasn't going to corrode. There could be compound curves; back when home craftsmen made their own bodies, the compound curve was impossible in metal. The shell itself could be made thick here for strength, thin there for lightness. Parts could be chopped off and glued back on.

Sounds great, eh? It was. Hall and Green and helpers constructed bucks for the various parts. Then they laid out the sections, or boxes, with fiber and wet resin, which they placed in a female mold and then cured in an oven. The sections were riveted together until the glue dried and cured.

The tub consisted of eleven boxes. It had a bathtublike floor and side sections, a center seat and firewall section, inner walls, a footbox forward of the cockpit and so forth. The engine bay was open at the back, to give access and allow for the gearbox.

Because fiberglass depends on mass for strength, all the direct loads, as in engine and suspension mounts and pickup points, were taken by metal bonded into the fiberglass. The engine bay tray was alloy and metal braces were located at strategic points.

The target weight for the tub with fixtures was 150 pounds. It came in at 144.

Doing the engineering and planning took time and knowledge. Beyond that the car took faith, first that it could be done and second that it was worth doing. Hall said years later that when they had the structure completed, when it was doing what they'd hoped it would do and when they knew the idea was going to work, it was perhaps the high point of the whole project.

Another point: Anglophiles in the audience may point out with some truth that Chapman's Lotus Elite, the first one, the lovely little coupe, was built in much the same way. Plus, the Elite *was* a monocoque, a coupe with the skin all the way around. Right. But the Elite literally came unglued after some time on the road. The Chaparral 2 held up and was repaired after crashes; a 2 that crashed in 1964 was glued together and won races in 1965 and then was converted into a coupe and went racing in Europe.

The fiberglass tub was a technical achievement that topped—and still tops—anything any other special builder managed to do. Beyond that, though, the fiberglass was overshadowed by techniques and victories much more entertaining to the lay audience.

We're not talking innovation for its own sake. The 2's suspension at first came nearly straight from Lotus. Not from the sports cars, which were too small, but from the Indianapolis cars that Lotus was building in partnership with . . . Ford. And where the soon-to-be pals with Chevy didn't have Lotus bits, they had Cooper bits, for instance the steering.

The suspension layout itself was conventional, independent at both ends. In front it was A-arms or wishbones with coil-over shocks and anti-roll bar. In back it was wishbones on the bottom, single link and two trailing arms on top, coil-overs and anti-roll bar. Brakes were 11.5 in. discs from Girling.

There should be a distinction here between early and late versions of the 2. The cars themselves, all three that were made during the model run, were unchanged in basics. But the Lotus and Cooper suspension parts were changed to uprights and hub carriers from Chaparral, the wheels were Lotus then Cooper then Chaparral's own design and on through the parts book.

Further, and against everything one thinks of when one thinks of Jim Hall and Chaparral, when the 2 was first laid out, there were two cars. One was built with an iron Chevrolet V-8 and four-speed Colotti gearbox, the other with a B-O-P V-8 and a transaxle made from Corvair parts and case. The B-O-P engine was an excellent unit, the Corvair kept the source in the family and anyway—witness the Indy Ford stuff—Hall wanted to solve problems and see what worked. He wasn't concerned much with the flying of flags.

The lack of test figures is a special loss here. Journalism goes through stages just as racing does,

A side view of the front torque boxes. Metal panels were bonded into the fiberglass at the suspension mounting and pickup points. Petersen Publishing Company

and the motoring magazines have times when the editor likes racing, the staff goes racing and they all test racing cars, followed by periods of drought or by periods during which, say *Road & Track* didn't track test and *Sports Car Graphic* did. It's also true that Hall was an introspective man, marching to his own drum and with his own interests, even secrets, to keep. He wasn't keen on letting strangers get too close to his effort.

Why the lack of inside scoops will come clear, but it is a shame because the car must have been incredibly fast.

Hall and Sharp continued to race the 1, while keeping their hands in with various Coopers and the like, during 1963. Late in that year Hall and the 2 showed up at the Canadian Grand Prix (for sports cars) but withdrew when there were too many bugs. At Riverside for the *Times* Grand Prix, Hall shattered

Some details needed to be worked out at the 2's first appearance, as in the cooling hose for the brakes. The collector exhaust pipes were routed out the rear, and the stacks on the carburetors were short for power at high rpm. Petersen Publishing Company

The new 2's flat rear was based on Kamm's chopped-tail aerodynamic theory, and had big outlets for the exhaust pipes. That's Hall on right, smoking a cigarette and facing away from the car. Petersen Publishing Company

For its second appearance, at Laguna Seca in 1963, the Chaparral 2 had ducts for cooling the front brakes cut into the front dam. Road & Track

the old qualifying record and stopped a few hearts when the new car demonstrated what looked like terminal problems with the front end lifting at speed. His crew cobbled up a spoiler, more like a metal goatee, and kept the front wheels on the pavement long enough to lead the race until the electrics caught fire.

The team had tested both the Chevy-Colotti and the B-O-P-Corvair drivetrains and the Chevy was obviously better and stronger, so that's the combination with which the 2 began its racing career.

Aerodynamically, the Chaparral's problem was one shared by virtually every racing vehicle of the time. This is strange in a way. Cars, especially Indy

```
Chaparral 2 (early version)
Wheelbase ............................. 88 in.
Length .................................... na
Tread ...................... 53 in. front, 51 in. rear
Weight ........................... 1,300 lb. dry
Engine ........................ Chevrolet iron V-8
Displacement ............................. na
Claimed power ........................ 365 bhp
Top speed ................................. na
0-60 mph .................................. na
Quarter-mile .............................. na
60-0 mph .................................. na
```

The public knew that the Chaparral had a secret gearbox, so as part of the game a crewman covered the mechanism in question...

...Because the crowd will stoop and/or crawl to see under the back of the car whenever they can. Pete Biro

An ultra-short fisheye lens allowed photographer Pete Biro to get this in-cockpit shot of Jim Hall at work in the Chaparral Two at Laguna Seca. (Note Biro's knees at lower left.) Pete Biro

One year later, at Mosport, Roger Penske lined up the Chaparral 2 between Bruce McLaren, in Penske's old Cooper, and Augie Pabst, who had finally replaced the old Scarab. Note the Chaparral's completely new nose. Road & Track

At Laguna Seca in 1964 the rear of the 2 was new, too. The flat panel had been replaced by a flat panel topped by a spoiler, the better to keep the rear wheels on the ground, while the intake stacks were much longer, for more punch off the bottom end. Road & Track

and Grand Prix style, had been going fast for a decade, but all of a sudden or perhaps because the engine was in the back and the arrow doesn't go straight like that, cars were sailing into the air. Phil Hill and Richie Ginther talk about their time at Ferrari and how they worked with the engineers installing lips and spoilers on the first of the rear-engine cars. Ginther says he suggested shapes he'd seen on jets when he was in the Air Force.

The Chaparral 1 benefited from advice given by GM technicians, and the first front on the 2 was mostly the shape evolved with the 1. But it didn't work. They all began to figure out why, with results that changed the shape of racing for years to come.

At Laguna Seca the 2 was third, behind a Cooper-Chevy and the last Scarab with B-O-P V-8. Then came Nassau, with more time in front and two more retirements.

People were paying attention. *Road & Track* seemed to be carefully impressed, within limits: "The basic Chaparral Two is a relatively conventional racing car in general layout: midships engine of large displacement driving through a transaxle to the rear wheels; unequal length suspension front and rear; large disc brakes and a light, structurally rigid chassis enclosed by an aerodynamic body shell." The magazine went on to talk about the fiberglass and why it was used, adding that aluminum would have been lighter.

What nobody knew was that changes were being made. At Nassau the Mecom team used aluminum-block Chevy engines, presumably part of the project in which Penske had been involved.

Tests between the first two 2s convinced the team that the Chevy engine and Colotti gearbox were the way to go, so that's how the cars ran their first races and how the magazines described and portrayed Hall's newest creation (Hall was by this time well-known as a builder and driver).

What they had was a strong, light car. So did the other guys. Mecom and Foyt and the incredible old Scarab, the rear-engine one running under the Zerex banner of all things, beat Gurney and his Lotus-Ford at Daytona, an event otherwise distinguished by stock car star Fireball Roberts and his 427 powered Fairlane sedan coming in second while Hall and others had mechanical problems.

What Hall and Sharp also had was an aluminum V-8, not for them alone mind, but one of a series that came from Chevrolet's experiment department. The aluminum engine wasn't as powerful as the cast-iron version; something about the high rate of heat transfer taking the heat from where it should have expanded the air and putting it where it boiled the water, just as in the B-O-P aluminum engine. But the new engine did weigh less and they had plenty of power.

They had so much extra power that they switched to a new transmission.

This is one of racing's great legends, so we have to spell it out, on tiptoe.

Perhaps the best punchline is that the team ran with the new transaxle for two races before anybody noticed. Jim Hall won Laguna Seca and Roger Penske, who drove for Hall when a car was available and he had nothing else to do, was third. At Kent, Hall retired and Hap Sharp was second. Then, at Mosport, Canada, which was won by John Surtees in a Lola, the Ford team realized that there had been something funny, something oddly regular, about the Chaparral's engine sound. Dave MacDonald said Hall wasn't taking his hands off the steering wheel.

Right. Hall wasn't shifting gears.

A lot of this was psychology. Back in the fifties there had been flocks of Allards with big V-8s and automatic transmissions. Much more recently there had been Bill Sadler's experiments with a single-speed gearbox. So the automatic transmission in Hall's car wasn't really new.

What it was, mostly, was secret. Hall was a natural introvert with a cool demeanor and an ironic view of the world. It suited his style perfectly to smile and say nothing, while the crew was instructed to keep a tarp over the rear of the car any time the body was off, driving reporters into even more of a frenzy.

This is Penske, relieving Sharp at Nassau late in 1964. The Chaparral's front had been changed again, with scoops for the brakes and heat extractors glued to the tops of the front fenders. Hall had help from GM in this aerodynamic study, *and this front looked a lot like the fronts of certain show and project cars seen at Chevrolet during the period. Road & Track*

The frenzy was matched by pique, from the press and from the other teams. Chevrolet wasn't in racing, as decreed by parent GM. Ford was very much in racing, with Carroll Shelby and others, in the United States and England, building prototypes and production sports cars and GT coupes in order to meet and defeat Ferrari and the others "Over There." Which they did—but that had little to do with the Ford puzzlement and distress, real or invented, over Hall's secret weapon.

After Mosport, Carroll Shelby sent a telegram to Frederic G. Donner, GM's board chairman, asking to buy "an automatic gearbox developed by Chevrolet Engineering Research and Development Division for use in the Chaparral sports car driven by Roger Penske which raced at Mosport in Toronto . . . it appears inconsistent to compete constantly with a firm which publicly contends to have no interest in racing."

Better read that again.

The despatch doesn't reflect well on Shelby, who on most occasions was the best-mannered of men and a good sport even when he lost. The telegram isn't a request or a question; it's an argument, a statement of several facts not in evidence, as they say in court.

He was answered, superbly, by Semon E. Knudsen, Chevrolet's general manager and a man who reportedly had been under the gun because of those alloy engines that Mecom raced in Nassau a few months earlier. Knudsen handled his part perfectly. He replied that "the unit in question is an adaptation of a production torque converter. Since you ingeniously adapted production parts to your vehicles and have considerable engineering available, this type of adaptation can easily be made by you. We have no units of this type available and the unit in question is the property of Mr. Hall."

Not a fraction of an inch from the factual truth. *Car and Driver*, in which this exchange was first reported, obviously in cahoots with Shelby, took offense: Who adapted the thing? they wanted to know, and even if Hall owned it, who gave it to him or sold it, or why?

What the magazine left out was, Who cares? How Chevrolet or Ford handled their involvement with engineering or experiments or racing was their busi-

A later version of the 2 was much neater, with canted radiator and headlights carried by a fabricated structure attached to the front box. Road & Track

ness. And it's fair for us to guess now that what really got up the press' collective nose was that they didn't know themselves who was giving what to whom. If there's one crime the press can't forgive, it's not being told the story.

And Hall wasn't talking.

Since then Paul Van Valkenburgh, who was a Chevrolet engineer assigned to such projects in the early sixties, and Phil Hill, who was world champion in Formula One and drove for Hall, and who is a literate and fair man, have told us more. Hall himself, while still not given to babbling details, will fill them in if the right questions are asked at the right time.

First, the fabled secret automatic transmission wasn't an automatic transmission, not according to the usual definition of a transmission that shifts for itself. The Chaparral never used such a thing. Instead, as Knudsen's reply stated and the magazine skipped over, early in 1964 the Chaparral began using a torque converter, what might be called a fluid clutch.

The torque converter was simply a hydraulic connection between the engine and the gearbox, which at first had the one forward speed. There was a shift, into gear and out of gear, by means of dog clutches. This was only to allow the car to be pushed back and forth or for the engine to be run while the car sat still. The impact of jamming the box into gear with the engine running was severe, so severe that the rule was Always engage gear before firing engine. (Hill comments on the care this required, while Van Valkenburgh in turn said that of all the drivers who worked the lever, Hill was the most mechanically sympathetic and did the least damage.) But once the box was in gear and the engine running, the work was done.

There was clear thinking here. When the 2 was under construction there was no doubt that the engine would have more power than the light car and small tires could use. There's no point in delivering to the rear wheels more power than the wheels can put on the ground. So, a properly tuned torque converter was set to deliver only as much power as could be used, until the car was going fast and the converter could lock solid.

The 2s ran this way for six races, while the team pondered the next step.

The rear of the revised car was also cleaner, with oil and brake cooling handled by means of ducts integrated with the rest of the car. The long, individual exhaust pipes would give more peak power but would narrow the optimum rev band of the engine, which surely means this car had the two-speed gearbox. Road & Track

The Chaparral 2 was as formidable in its own way—that of sheer competence and skill—as the Listers or Allards had been. Barely visible here are the extractor panels on the inner sides of the front fenders, the radiator air exit just aft of the number and the second row of louvers on the inner edges of the rear fenders. Chaparral and GM were just about to wonder why, if they could keep the air from doing harm, they couldn't make the air do good. Road & Track

That step was in effect a counter-step. The alloy engine had less power than the iron one, so there was less wheelspin to prevent, as it were. And the tire people were catching up. It had been proven by this time that the fatter tire did give more grip. Tires were getting fatter and softer and stickier, and the entry of other brands into racing didn't hurt that form of competition at all. But with less power to transmit and a better link to the ground, and with the competition improving all the time, one speed wasn't enough

The rivals aren't far behind. These are two old pals, Zerex and Foyt, but the car is a much-modified Cooper: note the airfoils just in front of the front wheels, the exit for radiator air on the front center panel, the exits on the trailing edges of the front fenders, the intakes on the front of the rear fenders, the louvers at the back and the rear spoiler. This car was also billed as the Hussein, in honor of the king by that name, and used a Dodge Hemi V-8 so big that the stock-car-style air cleaner sat above the bodywork just behind the driver. A fearsome weapon, but not nearly as good as all the parts should have made it. Road & Track

for the Chaparral. The transaxle was fitted with two speeds, still using the dog clutches, so the car could come off slower corners faster.

Remember that the larger and broader the power band of an engine, the fewer gearbox speeds are needed, or even used. Furthermore, when drivers were fat and tires were skinny, the brakes were weak and the engines were strong and racing techniques included downshifting—jamming the box into lower gears so engine braking would help slow the car. The disc brake and racing tire ended that. The brakes were so strong and the traction so high that cars could slow for corners in less time than drivers could shift, clutch, shift. Downshifting became a lost art and a waste of time.

While the other drivers were moving their left feet from brake pedal to clutch pedal and back again, the Chaparral drivers were learning how to effectively control the brake with the left foot and the throttle with the right. The early one-speed was especially good at this because the car's balance and the traction balance didn't change with downshifts, the way they do with the standard stick-controlled gearbox.

The two-speeds required more care, because the throttle had to be eased off so the gears weren't torqued against each other. Then the shift lever was moved and the shift performed by the driver, just as with conventional machines. (Later in the Chaparral story the box got three forward speeds, when the competition was tougher still, but that didn't come until after the 2s had been retired and the climate had been changed.)

There are two odd bits here. One was that the press had a grand time making a fuss over the secret and the technology, and of course the involvement or lack of such of Chevrolet and GM in racing. Hall says now that the secret was as much entertainment as anything else, while Knudsen's reply to attacks was factual. The gearbox and converter were Hall's property. If somebody else had developed the idea well, heck, if we didn't borrow things and techniques from each other all the cars in the world except for GM products would still start with hand cranks.

There is no statute of limitations on business gambits and most of the people who worked behind drawn blinds are retired by this writing. Even so, none of the people who were there are eager to name names, just as a matter of principle.

What they say in general is that there were people in Chevrolet and GM who knew how much sales had improved because Chevrolet and Pontiac and to some degree Oldsmobile had become performance cars, young men's machines, and they knew of the enthusiasm and creativity engendered by involvement with racing and with experimenting. Nor did it hurt that some of the engineering staff had practical experience working with alloy blocks and heads before the gas crunch and federal standards made such equipment economically viable.

At any rate, the men in the shops and at the drawing boards worked on what worked, while those higher in the organization took their own steps; a man who didn't dare know about certain projects might, for instance, let his staff know that he'd be coming into the shop tomorrow and when he arrived he wouldn't see anything he couldn't mention to his superiors.

Foolish? Sure. But the racing ban was just as foolish and sometimes the outcome does justify the act.

The connection

Sports car racing at the top had become a large and friendly family. Because the competition was as fair as it was fierce, the best drivers could and did race against each other without mercy, then swap mounts for the next weekend. Hall could build a car and sell it and race against it in another brand, while Penske could and did drive for himself, Mecom and Hall.

Hall was also a partner in the Midland, Texas, Chevrolet store, so he knew all the good numbers, and one day while still in the Chaparral 1 stage he dropped in on the Chevrolet Engineering office. He was an engineer too, and many of the staff members were racers so it was natural for them all to become friends and to appreciate each other's work... and to see how the two sides could fit together. Nothing formal just then, surely no plan for Hall to become Mr. Outside for all the insiders, just a way for clever minds to share ideas and parts. The Chevy guys were impressed, is the main thing here. There was no hint that they were going to help Hall because he needed them. More likely they were willing to work with Hall because he didn't need them, or anybody else.

What Hall needed was stuff like wind tunnel time and expertise, and that's what he got, while giving Chevrolet parts exposure and testing they'd never have had any other way.

The punchline was that the fiberglass-chassis Chaparral 2 came from Jim Hall, Hap Sharp, Andy Green and nobody else.

The details of the cars, usually two for each race meeting, varied with time and circumstance. The engines were always Chevrolet, iron then alloy block and heads. The first exhaust system was tuned—that is, the pipes were joined into connectors and routed horizontally out the rear body panel—then later the exhausts were stacks, one pipe for each cylinder, aimed nearly straight up. Most times the engines wore Weber carbs, four two-barrels, 48 mm then 58 mm venturis. The early engines were built by Art Oehrli and were 327s with the classic Chevy dimensions of 4 in. bore and 3 in. stroke, with Iskendarian roller tappet cams and 11:1 compression ratio and flowed ports and valves and combustion chambers. The engines were as good as anybody had then, but they weren't the secret of the team's success...

...If there was a secret other than hard work and careful planning and attention to detail.

Up from the ranks of racing fans came George Follmer, who began with a Volkswagen Bug and parlayed this Lotus 23 with a Porsche engine into the national professional driving championship, then a full career as a pro driver and as a successful businessman and dealership owner. Road & Track

By the racing record, the 2s began with one third place and three retirements in 1963. In 1964 when Hall won the USRRC at Pensacola, Florida, *Sports Car Graphic* said, as five cars beat the previous lap record, "We can once again see the evidence that the big American engine in the sophisticated British-type [Boo, hiss!] chassis has developed into a tremendously competitive machine."

Why the boo and hiss? Because the old record holder was Penske's Zerex Cooper, a true British-style chassis, while the new winner came from Texas and didn't copy anything. Used, yes. Took advantage of, sure. Improved, no question. But it wasn't a copy.

Back to the record. The competition was tough. Mecom's team cleaned up at Nassau with the lightweight Grand Sport Corvettes, Penske's Cooper-Chevy and all the gear the team could borrow from the "defunct" (quote from *Sports Car Graphic*) Chevrolet experimental department. That week was also marked by an incident that's hard to believe: Penske was leading a race until the engine overheated, so he came in for water and was so distraught that he zoomed out of the pits with a crewman still draped over the car. Yes. The Mr. Cool of Indy in a fit of pique. Sorry we don't have any photos and it's OK to laugh because the poor wretch wasn't hurt.

Back to the record. A. J. Foyt won at Nassau and Daytona with the good old Scarab-Chevy, a Genie won at Marlboro, a Cooper-Chev took Riverside when Jim Hall slid into the hay bales and Ralph Salyer was winning amateur races with the Cro-Sal Special, in fact a Cheetah with the top chopped off.

Then the Chaparrals got down to business, with Hall and Sharp, Hall and Penske, or Sharp and Penske ruling the racetracks. They were first and third at Laguna Seca; first and second at Watkins Glen and Meadowdale and Mid-Ohio; Penske won both heats at the second Laguna Seca meeting; they took the Governor's Trophy at Nassau. In the big race there Penske led until his car broke, so they called in Sharp, then Penske took over and won again, with both men getting credit for the victory. Hall was sidelined with a broken arm from a crash at Mosport, which is why they were doubling up.

By Watkins Glen time in 1946, the press was reduced to noting which of the two 2s had a clutch pedal (Penske's) and which had a gearshift lever (Hall's), while *Sports Car Graphic* commented that "the only question is which Chaparral driver will be USRRC champion."

It was Hall. Even with the trophy on the mantle, so to speak, he kept on racing and in the final USRRC of the year took the pole and led until the brakes failed.

It's true, the other guys were in disarray. Colin Chapman had introduced his version of the big howlers, the Lotus 30, but it was so bad that Parnelli Jones in a production Cooper-Ford lapped the legendary Jim Clark at Riverside. A. J. Foyt was doing a sponsorship deal; well, perhaps it was that John Mecom the oilman had a pal in the Mideast and saw fit to build a Cooper with Dodge NASCAR Hemi power. It was one of the first big-block V-8 specials, it got to wear the banner of Zerex, but it wasn't as strong or agile as it was fast and powerful, and Foyt had to drop out.

That last 1964 pro race was won by Walt Hansgen and Augie Pabst in a Ferrari prototype long-distance racer, just about the last endurance machine that would take the prize by outlasting the sprinters. Hansgen won the Bridgehampton Double 500 in Mecom's gallant old Scarab, while at the Canadian Grand Prix Bruce McLaren, who bought the old Zerex Cooper and took it to England for revision, ran a new and improved version, with a stronger frame, 15 in. wheels, a new body and a Traco-Olds engine. He called it a McLaren and said he planned to build his own cars from now on. More on that to come.

To make a success story short, in 1964 the Chaparral team made twenty-four starts, won seven races, and had six seconds, two thirds and ten retirements or crashes (the official team record never shows any crashes, by the way, just retirements). There are more finishes than starts because of that odd deal where Penske won and retired in the same race.

While Chaparral ruled the big leagues, *Sports Car Graphic* and the SCCA did a grand job with the runoffs, the race meeting in which all the top drivers from the amateur ranks got to run Riverside against their peers and collect tow money. A Dolphin-Abarth won H Modified, the Webster Special with Jerry Titus was D Modified winner and the big-bore class went to Bart Martin in a Cooper-Chevrolet. This was a continuation of something we've seen since the days of Poison Lil, the elderly Grand Prix/Indy Maserati that used to run the unlimited class in the fifties. As we all would do if we could, the better-heeled amateurs were buying last year's pro machines, and why not?

Back on the drawing boards, though, there was new thinking.

According to *Sports Car Graphic* once again, "Jim Hall has proved that his Chaparral was capable of carrying him to the USRRC title, and now, with Roger Penske driving, has showed the world that the same Chaparral, well driven, could put down the best field of international drivers and cars ever assembled, on one of the most demanding circuits [Laguna Seca] in the country."

That was a strange way to put it. Penske had surely proven himself by that time, and had won as many titles as Hall, against the odds and even with his own car. But never mind that, the Chaparral was the best.

Going on, *Sports Car Graphic* said, "One thing is certain. After this demonstration, all the competition is on the trail of a comparable torque converter unit."

Even more strange, that one. Not the magazine's guess, because anybody would have thought the same thing. By that time Titus could accurately report that the secret system was a three-stage torque converter, an automatic clutch, with two-speed gearbox. Most of the good cars had enough power to do away with all those speeds and all that shifting, while everybody surely knew or could see the advantage gained when you could devote more time and skill to braking than to jamming the box through the gears.

Except that nobody else came up with a system as good, and tires got better so there was more traction and thus more demand for power, leading to engines tuned to deliver added oomph in a narrow rev range—that is, back to multiple speeds. . . . When we look at the long view we see the clutch still with us twenty-five years later, while Ferrari is the only experimenter and has some sort of semi-automatic shifter in Formula One. (What nobody did predict, as always seems to happen, was active suspension.)

For 1965, more change. The USAC had dropped out of road racing and concentrated on Indy and the ovals, while the SCCA had become *the* road racing sanctioning body. There was the USRRC, in the spring

```
Chaparral 2 (late version)
Wheelbase .................................. 91 in.
Length ......................................... na
Tread ..................... 53 in. front, 58 in. rear
Weight ........................... 1,750 lb. trackside
Engine ...................... Chevrolet alloy V-8
Displacement
Claimed power ......................... 410 bhp
Top speed ........................................ na
0-60 mph ........................................ na
Quarter-mile .................................... na
60-0 mph ........................................ na
```

and summer, and there was a new series, a short one, with races in the United States and Canada, called the Canadian-American Challenge Cup, or Can-Am.

The first USRRC of 1965 was a double surprise. First, the pole went to Walt Hansgen in a Lola-Ford. Then came the Chaparrals, followed by a McKee with 427 (the *big-block* stock-car-racing Ford V-8), while Augie Pabst in the Chaparral 1 (with engine in front) was 1.4 seconds quicker than a new guy named George Follmer. Follmer's new car was one of those logical leaps, a Lotus 23 with Porsche engine.

In the race the faster three retired, Jim Hall while walking away from the field, and the overall winner was George Follmer, who was running the small class, under two liters. Second was Mike Hall (no relation) in the monster McKee.

That was a rare event, to be made more quirky still because the rules of the series never imagined such a thing, and paid overall points to the small cars the same as it did for the big cars.

But that's later in the year. Meanwhile, as Jim Hall pulled into the winner's circle at Sebring, *Road & Track* thought, "No one had doubted that his remarkable Chaparral was going to be the fastest car in the race." Which it was. What they had doubted was the car's reliability and its capacity for rain. As it worked out, Hall had the race in his pocket from the start, and had built up a cushion that let him relax when the rain did come.

The Chaparral team was on a roll that went beyond the normal roll. As *Sports Car Graphic* said a few months later, "The only question was by how much Hall would lead Sharp and the answer was 16 seconds."

Yes, they were that good. Hansgen provided competition when he wasn't breaking the Lola-Ford or the Scarab-Chevy or Cooper-Chevy under the strain of keeping up. McLaren's new McLaren-Elva, so called because he'd arranged to go into production under the banner and using the expertise of Elva, the English sports and racing car makers, was also good. The B-O-P engine was bored and stroked to 274 ci and had 340 bhp when propelling a 1,275 pound car, so it was close on the power-to-weight ratio. And by that time the B-O-P engine was reliable.

But the Chaparrals won, pro race after pro race. At Bridgehampton they lapped the rest of the field, which must have hurt. As an extra goad, Pabst in the ancient front-engine Chaparral 1 was as fast as the other guys in the back-motor cars.

Even when they lost—at Mosport for instance, where John Surtees and his Lola-Ford were second in the first heat and won the second after Jim Hall retired—nobody questioned that Hall had been the fastest. They lost, one could say, but they didn't get beaten.

There was no lack of competition. The Lola was coming good and Chapman introduced the Lotus 40, based on the 30 but with some revisions to allow installation of the larger, 351 ci Ford V-8. (In time the new Lotus would prove as poor as the earlier sports racer, in sharp contrast to the world-beating Grand Prix cars, and wags would say the Lotus 40 was the 30 plus ten more mistakes.)

There were fifty-five entries for thirty places on the grid at Watkins Glen, where Hall set the lap record and he and Sharp waltzed away from the field. On the way west the team visited Indianapolis and Hall lapped the oval in the sports car; just curious, he told the press, although in fact he was doing a test for himself and people back in the experiments department.

There was so much interest in the professional races that small firms like Genie and Lola and Lotus competed just to turn out enough cars for all the racers with sponsors (or perhaps trust funds). So many such cars were built that the FIA, which had gotten even with the United States a few years earlier by banning all the big engines from the classic events, now decreed recognition of what would be called Group 9, unlimited sports cars, a class that would exist because the SCCA and the Canadians and later the English had persisted and kept the big howlers in racing.

The quirk of 1965 surfaced with some force at the end of the season. Because Hall, Sharp and Penske were splitting their points and lost every now and then, and because Follmer and his Lotus-Porsche were the best combination in the U2 class, when all the USRRC races had been run Follmer and his small car were the titleholders. This wasn't what the SCCA had had in mind, and it promptly changed the rules for the next year.

Not that it really mattered. In the first ten months of 1965 the Chaparrals driven by Hall, Sharp, Ron Hissom and Bruce Jennings (the last two top outsiders signed on for the endurance races at Sebring and Road America) made thirty-four starts. They got sixteen wins—nearly batting .500—five seconds, two thirds, two crashes, one nonstart and five retirements, and a few twenty-seconds and fifteenths when the car was running wounded.

And then the era of specials came to an end.

Technically, by the rules of this treatise, the beginning of the end came when Hall and Chevrolet signed a formal research and development contract. Hall had built his own test facility, fittingly named Rattlesnake Raceway, out back of his shop in the industrial section of Midland, Texas, a locale famous for the "awl bidness" and for being just about totally industrial zoning with cactus.

Back when he was working on the first Chaparral, Hall had noted that if you don't drive at racing speeds all the time, you lose your edge. And he kept secrets and had ideas to be developed at his own pace, so it made perfect sense to build a little racetrack, fully instrumented, on his own land.

The Chevrolet experimenters were up there in the sometimes frozen north, in Michigan. GM had southern locations but they weren't always what those doing experiments needed, what with people hanging on the fences all the time. So Hall and Chevrolet made a deal, with Chevrolet to use the Rattlesnake test track and swap parts back and forth, Hall to collect rent and at the same time benefit from whatever the researchers learned.

They learned a lot. Hall and Chevrolet became partners in the sense that after trading parts—the alloy engine and first torque converter came out of a

A side view of Hall's final version of the plain 2, as raced in September 1965. By this time the suspension and wheels were all Chaparral work and the cars themselves were virtually unbeatable in what had become the major leagues of road racing, the Can-Am and USRRC. Road & Track

rear-engine project car the Chevrolet engineers built for themselves—Hall did become the conductor of Chevrolet's racing program.

In all respect, there was nothing wrong with this. The two groups cooperated, as equals.

But when they built the Chaparral 2B, a mule and test bed never seen in public now or then, and followed that with the 2C, an aerodynamically superior version of the 2 except that the tub was aluminum, the true originator and owner of the parts was the Chevrolet Engineering Department, experimental branch. Hall had become a sub rosa Shelby, and the Chaparral was a team car, just as the Super Cobras and GT40s were team cars. They were good cars, but they weren't specials and they weren't made by one man or a private team working on its own for the sport.

The dividing line here was picked by Jim Hall. At the time that he, Sharp and Green did the 2, Hall said, they hadn't done enough to make themselves valuable to Chevrolet. When Chaparrals ruled and did it using Chevrolet parts and research, the two teams were equal enough to be partners.

"It was a good relationship," Hall drawled in 1989, still not one to blab. And so it was, but when Sharp won Riverside in the 2 on October 31, 1965, it was the last major race won by a homemade car, a classic special in the spirit of Miles Collier, Briggs Cunningham, Lance Reventlow, Ken Miles or even Hap Sharp and Jim Hall.

It's hard to find a villain in this. The men and machines who've dominated sports car and road racing since have been just as smart, brave and dedicated as the amateurs ever were.

There's no villain.

There is a culprit.

Neil Young sings of "the aimless blade of science," which delivers facts and figures and data until there's nothing left to the imagination: "Nothing that they needed and nothing left to find," Young sang.

In racing, the aimless blade *was* science. The Chaparral-Chevrolet combination went on to invent myriad devices. They even took out patents. Giant corporations may be dull but the people in the shops and labs and even those staring out the window are often smart and clever and hungry. Collectively Chevrolet and Ford and Porsche and Mercedes-Benz and Ferrari and Jaguar simply had more resources and

Jim Hall, at the peak of his career as the best of the privateers and as the prime mover behind the best, most innovative and most influential special in the history of the sport. The back of this photo doesn't say what race Hall has just won; instead it says, simply, "Jim Hall. Big Winner." That's as good a summary as I can imagine. Road & Track

computer time and instrumentation and wind tunnel access than any private team could hope to muster.

And racing was too big to be left to the racers. Millions of dollars were at stake here, and when the purse is that important you can't wait for Richie Ginther to sketch a spoiler from aircraft memory or shorten the wheelbase because Bill Stroppe wants it that way.

Neil Young's aimless blade, not incidentally, "slashed the pearly gates." There are those, including me, who feel that way about racing science and the disappearance of backyard innovators.

Which isn't to say they all went home after Chevrolet came out of the closet. Rather, they went where they could still do some good.

Chapter 8

Back in the Old Backyard 1965–1970

The more things stay the same, the more they change

Ever wonder, when the embracing hero and heroine fade into credits, what became of the characters after the story was told?

Just in case...

Jim Hall and Roger Penske went on to become declared team managers for Chevrolet; later Penske bought or traded most of the industrial United States.

Briggs Cunningham retired from racing but had so many neat cars that he built a museum, with none other than John Burgess as head man. The museum was later sold to the Collier family.

Miles Collier died of polio in 1954. Ken Miles, Walt Hansgen and Jerry Titus were killed in racing accidents. Lance Reventlow died in a plane crash.

Max Balchowsky went into movie work, then into financial trading.

Bill Stroppe became a builder of off-road racing trucks and cars for Ford. He had to sell the Kurtis but a fan bought it back years later, and thirty years after his peers retired Stroppe is racing again, still with flathead Ford V-8 and still fast.

This preliminary update is presented here because with one or two exceptions the cast of characters, and the character of the cast, changed markedly in the mid- to late-sixties.

That's because racing changed.

One of many recurrent events in human history is what we call a return to basics. Jazz began as an

At the Daytona Speedway, 1965, the contenders line up for the SCCA's run-offs, the national championships for amateur drivers. As you can see, most of the amateurs have bought slightly used Can-Am cars, not long from the pro ranks. And one man has kept the faith, with a car he made at home. It's good to see that the spirit had not disappeared.
Road & Track

Just for the fun of it, here's another sort of aerodynamic experiment, the Chevrolet-powered PAM Special, circa 1966. It was a journeyman team of builders and drivers, but they took the theme of low windscreen as far as Lister did and they tried some aeronautical stuff with the airfoil/control panel/buttress combined with the roll bar. The car didn't do much, though. Road & Track

elemental alternative to more formal music, for instance. But when Dixie had become bebop, a cerebral exercise that was so complicated only the musicians themselves knew what they were doing, we invented rock 'n' roll.

Thus, when road racing became an exercise in science by degreed teams of engineers dealing with budgets beyond cut-and-try, the SCCA of all organizations came up with some alternatives. One was a fair system with which to compete for national championships in the amateur ranks, with production-based cars that a clever person could improve better than a less-clever person could.

Better box office was the invention of the Trans-Am, a professional series for racing sedans. Well, that's what the rule book called them. They were really pony cars, Mustangs and peers. They were short-coupled coupes with small-block V-8s: each major manufacturer had one. The engines and suspensions and unit bodies lent themselves to improvement. At the top were factory teams operated by guys like Penske and Hall and Gurney, while in the pack were new teams and talents, men who could get a toehold and a reputation with a well-built production-based sedan.

The Can-Am grew while the USRRC shrank. This was good in that prize money and public appreciation increased and it was good in that more people had a shot at winning. In 1965 the Chaparrals won eight of the nine USRRC races, but in 1966 three different drivers and teams were in front for the first three races. Some of this came from the departure of Chaparral to contest FIA races in Europe, along with Ford, but some came because all the cars got better as the competition got fiercer.

Hall had applied for a constructor's license, which allowed him to enter FIA events as a brand of car. Not for many more years would Chevrolet resume open and outright participation in racing.

The innovations continued. The 2C, the roadsters with aluminum tubs instead of plastic, also had improved aerodynamics in the front, and a variable

This is an early version of the all-conquering McLaren-Elva. McLaren was a good Grand Prix driver who was a better planner and manager. He reasoned that the best conventional car, with optimum engines and driving, would in the long run prevail in the pro series, and he was right. Almost-as-good customer versions of the McLaren were built by Elva and sold to private teams and filled the ranks. Road & Track

(from the cockpit) rear spoiler, a huge flap that leveled out down the straights and jammed the rear wheels onto the ground under braking. Having only two pedals for engine and brake was a help here, or so the magazines said until someone pointed out that raising or lowering the rear flap was something the driver would do only when he needed to control only one other pedal anyway. At the time this was mentioned, the conclusion was that Chaparral had invented the two-pedal advantage and the others had been so snowed by Hall and crew that they'd accepted his reason why they couldn't copy him. And it may have been just that way.

The best possible illustration of Chaparral influence and creativity and cooperation came when the team astounded the racing world. At Bridgehampton, Can-Am East, in late 1966 the Chaparrals came off the trailer wearing "slab-like airfoils that towered above the bodywork on two vertical posts," said *Road & Track*, adding that the cars looked "like a pair of 21st century golf carts."

"The SCCA could find nothing in the rules to make the Chaparrals illegal," continued *Road & Track*, "so they ran merrily—and mysteriously—on."

The reverse wing came about after the earlier front mustaches or spoilers, which were followed by fixed lips to keep the rear of the car on the ground and then by the adjustable rear flaps. The high wing fed directly to the rear wheels—versus the chassis—and didn't affect the suspension at all. It was so new that a patent was issued, in the name of Hall and some of the Chevrolet men who used to hang around Rattlesnake Raceway. Seems they'd done so much mutual bench racing and concept sharing that nobody could remember whose idea the high wing had been, so they all got the credit.

The Chaparrals didn't always win, though. Chaparral had long since abandoned any idea of building customer cars, while in the first three 1966 races there were eighteen different McLarens in private hands.

The winner of the Can-Am title in 1966 would collect $50,000, from a total purse of one third of a million dollars. The winner of Le Mans got $5,000, period.

Later, the full field of McLarens became Team McLaren and then "The Bruce and Denny Show." McLaren and teammate and fellow Grand Prix driver Dennis Hulme did the best job of working out the aerodynamics and fat tires, and getting the most from the aluminum big-block Chevrolet V-8s, known as Rat Motors because the smaller engines were Mouse Motors.

McLaren owned the Can-Am, despite Ford and Chaparral. The latter kept coming out with new and different ideas, culminating in the 2J "sucker car," with enclosed bodywork and powered extraction of the air from beneath the car, resulting in traction beyond anything then seen or even believed possible. There were mechanical problems and naturally legal problems.

At one point I was covering the races for *Road & Track* and asked Hall, Why all the freak machines? Easy, said Hall, because we can't build a McLaren any better than they can, we'll have to do something else. Of course, I said, and I thought of John Cooper and his little cars with the engines in back.

When the legal and technical struggles were done the McLarens and the big-block alloy Chevrolets were the best in the Can-Ams of their day, and they ruled the roost until the next day arrived.

Bobsy SR3

Specials were clearly in eclipse. In the 1965 ARRC run-offs, in which the winners were officially declared SCCA national champions as well as run-off winners, only one champ could have been described as homemade. And that was because the professional builder of the car was its amateur driver.

Driver-builder Jerry Mong liked to work things out for himself. He'd been building Bobsys for several years at that point and he'd kept abreast of what was being done.

On the occasion of his winning car being analyzed, he told *Sports Car Graphic* that the fuss over superlight cars was nonsense.

The Bobsy SR-3, driven here by builder Jerry Mong at Daytona, was built heavier than the earlier Bobsys because Mong knew the stickier tires would produce more cornering power and thus more stress. Road & Track

The last American hold-out

Dick Durant was an aircraft engineer from Florissant, Missouri, a suburb of St. Louis. He had a lot of guts and a lot of brains, but he didn't have much money and his luck was all bad.

In this book, Durant is a hero. He raced in the Midwest division, in his own special. It was a truly homemade car. The body was Devin, with aerodynamic modifications and additions, and because Durant's interests were strictly in performance, the tacky old fiberglass shell made one think the car was on its last legs.

Wrong. Durant thought like an engineer. When he decided to make his own car, he knew what he could and couldn't afford. So the car would be front engine, with Corvette power.

Then he designed his own space frame. But before he built it with metal in full size, he made model after model of frames, tiny ones in match sticks. He had a strain gauge to scale and he tested, over and over, where the frame needed to be stiffer and stronger and where it could be lighter. Before he did the real one, he knew he had a design that would work.

The design was right and the Durant Special was a winner. True, the Midwest division wasn't a hotbed of late-model sports racers in class C. But also true, many of the tracks in the division were small and informal; for example, the public parks in various Kansas and Oklahoma towns were less-than-perfect places to unleash the classic huge engine in a small car.

The upshot was that Durant and his obviously outmoded beast qualified for the 1965 American Road Race of Champions (ARRC), the run-offs as we call them now, which was held at Daytona Beach under the auspices of *Sports Car Graphic* and the SCCA.

This sounds like a fairy tale, but the Durant Special had one tiny weak point: It used a Chevrolet engine, with fuel injection from the optioned Corvette. It did that because the units were easy to find back then, and because Durant could get more power per dollar than with anything else, for example a set of Webers.

Drive for the pump for the fuel injection was a sheathed cable, the sort of thing you see mostly driving tachometers and other low-stress items. Why Chevrolet used the cable, nobody ever explained.

It probably worked on the street. On the track at full revs, however, the speed and power sometimes proved too much for the little metal strands and they parted.

Yep, there was Dick Durant in his homemade car, lined up against all these people with modern cars, fresh from the factories. The other drivers had rear engines, more power and less frontal area . . .

Dick Durant took the lead, the amateur version of Parnelli Jones, with heart and guts getting more from the obsolete warhorse than even the builder believed was still there.

Except that Jones won. Durant was on his way to winning when the injection pump drive broke and the last holdout, the last competitive front-engine special, coasted to a stop.

The last front-engine special is in the lead, as Dick Durant powers his homemade Devin-bodied Chevrolet-powered car through a turn at Daytona, 1965. Durant was in the lead when a dumb little part, the injector drive cable, broke and cost him the race and the championship. Road & Track

Bobsy SR3	
Wheelbase	88.5 in.
Length	na
Tread	49 in. front, 49 in. rear
Weight	725 lb. trackside
Engine	Saab
Displacement	45 ci
Claimed power	72 bhp
Top speed	na
0–60 mph	na
Quarter-mile	na
60–0 mph	na

What mattered was cornering power, Mong said, and because we then had sticky tires, and fat tires, the cornering loads were more critical and so was the chassis' behavior under the extra stress. Thus, a car that weighed 500 pounds and flexed wasn't as good as a car that didn't flex, even if the stiffer car weighed 700 pounds.

The Bobsy in question, the SR3, weighed 725 pounds. It was heavier than the rivals because Mong had designed it with the new tires in mind.

He began with a space frame, of aluminum (recall the previous Mong efforts and views about that) tubing. It was heliarc welded and fully triangulated—that is, all the corners were triangles, the strongest shape in nature. With all brackets and hangers the frame weighed 47 pounds.

Front suspension was upper and lower control arms, with the attachment and pickup points sited so the arms were subject only to direct force—compression or tension—and not twisting or side loads. Rear suspension was wide-based control arms, upper and lower, using Bobsy's own cast uprights, with the coil spring-shocks feeding into the fabricated hoop that served as the back cross-member and located the transaxle.

Mong's own car ran in H Modified and had a Saab engine with three cylinders, 750 cc and 72 bhp. It was a two-stroke, with reworked port timing and tuned intake and exhaust. The engine was mostly stock, Mong said, because it didn't need any more power than it had. As is supposed to happen but doesn't always, the Bobsy-Saab was reliable because it wasn't fully stressed.

The SR3 might have been built with no regard for weight because of the extra stress from the new tires, or because Mong used the same basic frame and suspension for the middleweight modified classes, with engines up to two liters. The frame tubes had thicker walls, some suspension pieces were steel instead of aluminum and the front bulkhead was beefier, but the design was the same.

Mong did most of his business with racers who already knew the basics and could buy the kit, either in stripped form with frame, suspension and body, or complete except for engine but unassembled. He said he made it clear to prospects that they'd better have built other machines before tackling this one, and by the evidence the buyers followed his advice because the brand was in circulation for several years.

Competition was certainly not spotty in the big leagues. Ford's GTs were up against the Chaparral coupes, at least one of which was a good ol' plastic tub, a 2, rebodied as a 2D. And one of the 2s that crashed in 1964 was winning races, glued back together, in 1965. (What do you suppose could have been done with the material and techniques if GM or Chevrolet Engineering had *really* been experimenting?)

At the other end of the spectrum was another exception. At the 1965 run-offs where Jerry Mong was the only winner who'd built his own car and Dick Durant was the last front-engine special in contention and even Ralph Salyer, who'd raced the topless Cheetah, turned up with a McLaren-Elva, the man who won D Modified did so by prodifying his Triumph TR-4 until it was illegal for production racing and could race in the modified class. So weak or fragile was the opposition in the class that he took home the national title.

As perhaps another reflection, Chuck Parsons, who was a good but not great driver, was USRRC champion for 1965, equipped with a Chevy-powered McLaren-Elva. The McLarens were good, solid cars, better than the others from England and better than anything being built or raced in the United States. The disclaimer comes because Ford was winning world championships with GTs and prototypes, and Indy and Grand Prix with their engines and at the stock car tracks. Various sports racers showed up with some spin-offs, such as a twin-cam conversion of the truck V-8 with three valves per cylinder and the two cams one atop the other in the block's vee, not atop the heads. In general, however, it was Ford versus Chaparral versus the world overseas and Mustang versus Camaro at home, with McLaren and special project Chevrolet engines at the top of the sports car league.

Mirage

Just because the McLaren operation was tough doesn't mean it wasn't challenged.

One of the best private efforts came from Jack Nethercutt, a California sportsman who'd owned a Lotus 19 with V-8 power and given rides to top drivers like Dan Gurney. When the 19 was no longer in contention and the Chaparrals were at their peak, Nethercutt decided his team could build a competitive car, too. According to *Sports Car Graphic*, though, while Jim Hall had some anonymous help, Jack Nethercutt chose to do the entire project with his own people and no outsiders.

He made all the right moves. For instance, he hired an aerospace engineer who'd earlier designed and built a successful formula car; and a mechanic and fabricator who'd taught himself how to work with sheet alloy, as seen in the monocoque style of construction; and designer-aerodynamicist Pete Brock, who'd already done the body for the world champion Cobra Daytona coupe.

The team began with the newest, a stressed-skin, by-our-terms monocoque, of aluminum sheet riveted into an assembly of torque boxes, much as the Chaparrals had been done. The main boxes for the Mirage, though, were the fuel tanks on each side of the cockpit. Then came the tray below the occupants, the bulkheads front and rear, the boxes in front for controls and radiators, and the engine bay.

The last time the subject was chassis stiffness, the claim was 3,000 pounds of twist for every degree of deflection. Nethercutt told *Sports Car Graphic* that his monocoque frame resisted to the tune of 5,000 pounds per degree. Yes, this was the sort of claim one couldn't prove without damage to the structure, but it's worth noting that articles about the Ford racing cars—the world-class, factory-backed setups—claimed an incredible 10,000 pounds per degree.

Meanwhile the Nethercutt team went mostly conventional for the suspension. It was independent, with shortish control arms, anti-roll bars, coil spring-shocks set at an angle, cast uprights so the suspension links could be well inboard of the wheel itself and rack-and-pinion steering. Departures included wheels made of sheet alloy, in halves that were riveted together (an idea that startled people but worked and became something seen on other cars later), and only 13 in. tall.

That question has been raised before. By 1966, which is when the Mirage was first raced, the usefulness of the bigger tires was unknown. Against that, one had to balance the smaller frontal area of the lower wheels and tires, and the gain in top speed you got with the lower car, and the limit to brake size that came with the smaller-diameter wheel.

Nethercutt opted for the reduction in frontal area and went with the 13 in. wheel, and with brake discs that in theory would run cool despite their size. They were vented discs, which was new then, and the bodywork was designed to help extract heat.

The body was simply lovely. Brock began with scale models in clay, his thought being that you

Under the skin, the Mirage was up to date, with the central tub for the cockpit, engine in the rear and so on. All the components were as good as money and time could make them. Pete Brock

couldn't tell enough from two-dimensional drawings or little models.

His next theory was that wind tunnels weren't the last word. First, he said, what you get depends on the operator's understanding as much as the plain figures, and there weren't that many operators who understood racing cars and racing speeds. Next, a model sitting on a surface isn't the same as a full-size vehicle rushing across a surface that sits there.

For a clincher, Brock pointed out that a road racing car is seldom pointed straight ahead. It's zooming through turns or pitched into a slide or tipped back under power or forward under braking. Thus, all the data collected on a model or even a full-size shape sitting in a stream of air was not quite going to be the same as the data on that shape out in the world, on a track, at speed.

Brock sat down with the dimensions of the car and drew a body that looked right and that aligned with all the knowledge he'd gained doing work for the racing Cobras and others. The shape was lofted and expanded and produced in flawless fiberglass.

It was beautiful. And when they took it out on the track, the little tufts of wool that show where and how the air is flowing over the surface just sat there. The shape was right, and it was slick, and the car didn't fly. The shape was done to a principle, that of being the lowest possible drag.

The drivetrain was done in the same way, a search for the optimum power-to-weight ratio. The engine was a B-O-P V-8, in Olds cylinder head form. It was built by Traco, was bored to 280 ci, and with roller cam and four 48 mm Webers was rated at 375 bhp. The transmission was a four-speed Colotti. *Sports Car Graphic* asked Nethercutt about an automatic, as in Chaparral, and he just smiled.

The Mirage had a dry weight of 1,280 pounds, and with a full load of 34 gallons of gas and a driver it hit the grid at about 1,600 pounds, so the power-to-weight ratio was fine.

On paper the Mirage looked good and had all the credentials. It arrived at the track... and in headline terms was never heard of again.

The reasons contain two morals.

The first moral comes from Ford. The GT racing program was underway at about the same time, and one magazine reported that while the Ford GT's steel tub weighed 300 pounds, its torsional resistance was 10,000 pounds per degree. The sort of wind tunnel research Brock couldn't do for Mirage was being done for Ford. The Ford researchers didn't just know the drag and the lift front and rear at all achievable speeds, they knew drag and lift at a yaw, which is when the car is skewed in relation to direction of travel, as in a side wind or a four-wheel drift, up to 15 degrees. What even the best privateer couldn't afford, or perhaps wait to do, the corporation could do—once the board of directors or Henry Ford himself or whoever had decreed that it would be done.

The second moral involves a huge change in technique and application, one important enough to rank as a revolution.

Brock: "When we began on the Mirage, the key was drag. When we got to the track, the key had become downforce."

As the stock-block engines became larger and more powerful, and as the rear engine took over and put the weight at the back of the arrow, all racing

The trouble with the Mirage was that it was so slick and clean and streamlined. Brock's design supposed that cutting through the air was the important part. But between the drawing table and the starting line, there was a switch. The big new engines had so much power that they could power cars that weren't slick, that rammed through the air with fins and wings and spoilers that forced them against the ground. The beautiful car wasn't competitive any more. Racing is an odd sport. Pete Brock

designers began to deal with aerodynamics. As seen earlier, Hall and company were in the lead, with the front plows and rear flaps and vents and so on.

They had found something nobody suspected was there. At first the shape helped the car go fast, the low drag Brock cited. Then, curing the tendency to fly forced the designers to apply air pressure to the front, and when that got the back loose they had to force down the back.

Until one day some of them (not everybody reached this conclusion at the same time) said Eureka! If the car is being held on the track by air pressure, *we can go around corners faster!*

And darned if they didn't do just that.

There were still limits. But because there was only so much forward traction, so much tire grip to put power on the ground, you could build a car with the big-block V-8 Rat Motor from Ford or Chevy; both made aluminum versions while Chrysler had abandoned its big racing V-8s and the alloy Hemis were strictly for drag racing.

Ford had an alloy 427, Chevy began with a lightweight version of the truck-born 396 that then went stock car racing as the Porcupine engine. These engines cranked out 500 bhp, while splitting in half in the early versions. They became stronger and grew into 400 ci, 600 bhp monsters.

Surplus power is what Brock says they had. And they used that surplus power to propel the car to 200 mph while at the same time pushing it against the ground. The Mirage and the other sports racers with the small engines had the power to go the same speed only with slick aerodynamics.

The mix of traction, power and weight meant that the big and the little cars had the same acceleration, and the same top speed. Thing was, the big cars like the McLarens and Chaparrals and Lolas were through the turns quicker, thanks to their ungainly wings and plows and spoilers.

There had been another revolution, and, just as with front versus rear engine, those who hadn't been in the right place and taken the right notes were left out.

Progress, or at least change, works in leaps and bounds. When the aluminum chassis Chaparral appeared, it was known to be lighter but the principals made it plain that the plastic version damped vibration and the alloy one shook the driver's teeth loose. The plastic insulated the driver from heat, the alloy passed it on. And there were murmurs that Chaparral would begin production of a road car, an exotic GT sort of car with fiberglass chassis. Instead, the brand remained private and the material remained aluminum and nobody wanted to tell us why.

The specials were in low ebb. The 1966 Can-Am, the season that saw the Mirage not do anything of note, was won by John Surtees in a Lola, with Bruce McLaren and his own car second. The first nonfactory car (i.e., the first car that was not a Lola, McLaren or Chaparral) was a Genie-Ford. In the ARRC run-offs, C Modified went to Ralph Salyer with a McKee; H Modified was won by Don Parkinson and a Dolphin-Abarth. Allan Lader and a Cosworth-powered car built by a northern Californian, Borgeault, won G Modified, and that's all the homemades in the results.

By 1967 Mark Donohue and Roger Penske's Lola-Chevrolet, with strong Sunoco backing, cleaned up in the fading USRRC. Then Bruce McLaren arrived with a new, monocoque car and took over the Can-Am. The new car weighed 1,700 pounds, squeezed 500 bhp from its 358 ci, had to be wound to 7000 rpm to get that power and was tuned to such a narrow band of useful engine speed that it ran a five-speed Hewland transaxle.

So much for the single and two-speed gearboxes of the future. McKee was busy doing things like a turbine-powered car for Howmet, and he'd gone to Hewlands for gearboxes because it was easier than making his own. A fully prepared and race-ready McLaren went out the door for $40,000.

There was better news in the SCCA's smallest amateur class, H Modified. Ted West, a contributor to *Road & Track*, did an informal survey in mid 1967 and came up with an impressive collection of small (850 cc limit) one-off or very limited production cars. The joke is that Hm stands for Homemade, West wrote. But in fact at the run-offs at Daytona Beach the big cars, the Can-Am-size V-8s, averaged 98 mph, while the better H cars averaged 88 mph, with a fraction of the power, never mind the money invested. Here's a selection.

Kangaroo-Fiat

Fred Plotkin figured in the beginning of West's review because Plotkin used mostly hand tools and his backyard shop to build his own monocoque chassis, of aluminum, held together by rivets. He made a fiberglass body and adapted suspension and drum brakes from an NSU mini-car. The Kangaroo probably was named that because Plotkin used the original Fiat transaxle and engine, with the engine projecting aft of the rear wheels just as it did in the sedan from which it was taken. This gave a static weight balance of forty percent front, sixty percent rear.

The engine was a pushrod (overhead-valve) 850 Fiat, with raised compression ratio and two SU carburetors. The SUs were not the most impressive carburetors on the market, but they were inexpensive because people with more money to spend threw them away.

West gave no figures for this car except that it was second in regional points in 1966 and had won four of five starts by mid 1967.

LeGrand-PBS

Double introduction here. LeGrand was a builder of small racing cars, most of them single-seat formula cars. Late in their career, though, the makers

The Kangaroo-Fiat used a Fiat engine sticking out behind the rear wheels, just as it did with the sedan from which the special's builder got it. This put lots of weight in back but didn't hurt handling as much as you'd expect. Road & Track

branched into modifieds, or sports racers. Mostly these were sold as kits, for the engine of choice.

PBS was two brothers, Paul and Bob Swenson, who raced as kids but literally grew out of their go-karts. They were in the engine business mostly, and began with tuning Fiats. They got so comfortable at modifying the little Italian engines that they wound up building aluminum cylinder heads, with twin cams and hemispherical combustion chambers, for the Fiat 600 block. Said block and crank were bored and stroked to the 850 cc limit and with Weber carburetors claimed power was 85 bhp. The Fiat-based engines from PBS and Abarth and other tuners generally were as powerful as the converted English Fords if not always as sturdy.

One LeGrand-PBS had a space frame of steel tubing, Airheart disc brakes on all corners and a fiberglass body that looked a lot like the old Cooper sports car's.

Dolphin-Abarth

By the statistics, the Dolphin-Abarth was the best in class in the United States. Driven by Don Parkinson, it won the H Modified national championship in 1965 and 1966.

Dolphin was based in San Diego and specialized in the smaller classes, with some exceptions like the earlier Porsche-powered Dolphin special. This car used a space frame of steel—the monocoque was just then becoming the sort of project the average small

The LeGrand-PBS was mostly an expanded version of the firm's formula car. It was powered by a highly modified Fiat engine and covered with a fiberglass body that looked a lot like the old Cooper sports car's. Road & Track

The Dolphin-Abarth used the same body as seen in the earlier and larger model, of fiberglass and well done. The smaller engine was Fiat-based and modified by Italian tuner, Carlo Abarth. Parkinson was a good driver and the combination was tops in class at the time. Road & Track

builder was even willing to think about. It had suspension from Dolphin and drum (surprise) brakes from a small Fiat.

The engine was an 850, based on the Fiat 600 but enlarged and fitted with a twin-cam head from Abarth. The transaxle was Fiat 600 but the engine and transmission were swapped in the chassis; that is, the engine was in front of the rear wheels.

Saab Special

Tom Evans' Saab Special was something completely different, as the English comics used to say, or perhaps it was something from the past. The engine was a Saab GT, with porting and exhaust tuning, no power claim given. It sat in front, as it did in its original home. But it was joined with a BMC transmission, as from a Sprite, and then to a driveshaft and live rear axle. Front suspension and disc brakes came from an NSU mini-car and the wheels and engine were held in place with a steel tubing space frame. The aluminum body was hammered at home. Not modern for its time, more like out of date, but because it was light and basic and reliable, the car was competitive in local events.

West cheerfully and accurately reported that the small classes were still the best place for the person with a limited budget, and to an extent a limited ambition, because the drivers who were intent on going pro didn't display their talent in their own, homemade cars. The best self-made man from this time, George Follmer, began with a Volkswagen sedan in solo events; when he moved into the big league, he had a Lotus-Porsche (no small investment) and he had a good professional mechanic.

Eventually, Ferrari sent a few prototypes into Can-Am competition. About the same time Hall and Ford were sending their cars to Europe and Hall and Shelby were talking limited-production GT cars.

Builder Tom Evans made this body and frame by hand and worked out how to drive the rear wheels with a front-mounted Saab engine adapted to a BMC gearbox. Note the lay-down driving position. Road & Track

Racing was big business. In the fall of 1968 the SCCA created separate club and professional racing divisions, on the reasonable notion that the two venues had different goals and different needs.

The federal safety act of 1966 mandated a bunch of changes to production (road) cars. Some of the requirements were needed, more of them were simply ordering the car companies to do what they were already doing. The main point of the campaign was political, to take control of the motorcar from the sellers and buyers and give it to the government, the folks who made our post office what it is today.

The secondary and finally fatal result was that the small, specialized builder—the Kurtis or Allard or Devin or Reventlow—of tomorrow will not appear. Certification of a car became so costly, never mind insurance, that limited production ceased, cosmetic variations aside.

Back in the major leagues, the Can-Am was ruled by the Rat Motor, with 600 bhp driving wheels that were 15 in. wide. And in 1969, the final results for the season showed that only one man, Joe Leonard in a McKee-Oldsmobile, had earned a point in a car that wasn't series production or from a major team.

Quasar

By 1970 you needed a technically contemporary car if you wanted to take home any trophies.

Fred Puhn took home lots of trophies. He was an engineer, a member of a design group in aerospace when that field was where the smart people were. He went into business on his own, making racing wheels of spun—not cast—aluminum and magnesium. From there he designed and built his own car.

The little class, H Modified and later H sports-racing, was still the place for builders with more time and skill and training than money. Even so, Puhn spared little in the way of material or technique.

His car, named the Quasar for a mysterious object or force in outer space, began with a central hull of stressed aluminum panels, braced with tubes and bulkheads. In front of and behind the hull were subframes of square aluminum tubing, which fed chassis loads into the hull and located the engine and suspension. The body panels, more like covers for the tubing, were fiberglass, as were the two rectangular panels at the sides of the hull—panels only the FIA would recognize as doors.

Puhn did some creative thinking for the structure. But when he did the suspension and its geometry, he figured he didn't have the time to compute all the combinations. However, he knew that Brabham, the formula car builders, had done that, and that he was building a car of the same size as the Formula Three Brabham . . . so he duplicated the Brabham's geometry, with upper and lower A-arms in front, reversed A-arms and trailing links in back. Rear springs were coils and the front spring, singular, was a torsion bar (just like an anti-roll bar except the center was anchored in the frame).

Steering was Triumph. Fred Puhn did some reworking here. First, he provided an extremely tight turning circle. It was his and Colin Chapman's belief that the farther one could turn the front wheels, the more extreme a situation one could steer oneself out of. So the Quasar's steering lock went so far that with the wheels fully over in either direction, they were so turned that it was difficult to push the car forward.

Next, Puhn set the steering ratio to provide four turns lock to lock. That sounds foolish. Pioneer sports and performance types held most firmly to the tenent that big dumb cars had slow steering and quick little

Quasar took the low windscreen to its ultimate—the same height as the fenders. These cars were beautifully crafted and slick. The driver here is this book's author, who found the Quasar to be a better car than he was driver. Road & Track

cars had fast steering and the way you could tell was the fewer turns, the better. Except that Puhn had provided all that lock.

Finally, Puhn thought that giving more degrees of steering wheel turn per degrees of steered wheels turn made the steering more precise. His clincher was that road courses were fast and artificial and he'd yet to find one with a corner so tight that he had to take his hands off the wheel.

Brakes began with Airheart calipers and custom rotors, sized so they could be tucked inside the small wheels: 10x7 in. in the front, 13x9 or 13x10 in. in the rear. The height of the tire determined the height of the car, and the lower the car the less frontal area and the more speed, so Puhn built his own wheels as small as the available tires would permit.

The engine in Puhn's car seemed like almost an afterthought. At that time Chrysler had an English subsidiary called Hillman and Hillman made a subcompact named the Imp and the Imp had a good little inline four-cylinder engine with the right displacement, an overhead cam, an alloy block and a built-in tilt of 45 degrees, the better to reduce frontal area. It was a sturdy engine, well oversquare with the destroke needed to bring it within the class D 850 cc limit. With two Weber carbs and the usual modifications, the Quasar's Imp showed 88 bhp on the engine dynamometer. The gearbox was a Hewland, with six forward speeds in keeping with the small engine.

The Quasar was an impressive car, unbeaten on the West Coast in 1970, and it was the subject of a track test by *Road & Track*. After years of nonfigure testing, the results were recorded.

Quasar-Imp	
Wheelbase	92 in.
Length	na
Tread	55.3 in. front, 51.2 in. rear
Weight	820 lb. curb
Engine	Hillman Imp
Displacement	51.4 ci
Observed power	88 bhp
Top speed	124 mph
0–60 mph	9.4 sec.
Quarter-mile	15.3 sec. e.t.
60–0 mph	na

Good work, eh? Quasar duplicated the Brabham formula car's suspension geometry, with extremely generous steering lock. This is 10 pounds of car in an 8 pound package, but it all worked and it wasn't cramped to drive. Road & Track

Again, the Quasar is in such perfect proportion that the scale isn't apparent. The driver looks huge: he was not. One advantage of the little cars was that they didn't have enough power to utilize all those funny wings and things, and could be as slick as the wind tunnel and the designer could make them. Road & Track

As an unfair comparison, the Kurtis-Hudson tested twenty-five years earlier had a better time to 60 mph, but would have lost in the drag race and on the top end, never mind what would have happened in a road race. Just as true, this little Quasar was an impressive piece of work.

Subjectively, it was even better. I was the reporter who did the track test of the Quasar for *Road & Track* in 1970. At the time I was racing my own production sports car and track testing various road race sedans of the Mustang-Camaro-Barracuda persuasion and had done midgets and sprint cars and the like. The Quasar was the first contemporary, competitive, little purebred racing car I'd ever driven . . . and I wasn't prepared for it.

First, I persistently underused the car. I put on the brakes too soon and slowed for the corners too much; I wasn't ready to believe a car could stop that quickly or corner at such speeds.

Next (and more important in the long run), I overdrove. As a graduate of the pitch-it-and-punch-it school of race car driving, I expected to throw the car into lurid slides and spend most of my time with the car at right angles to the direction of travel.

The production-based cars of the day had to be forced into acting like racing cars. Not the Quasar. Not the really racing design with sticky tires and scientific suspension and proper balance. The new-model sports racers needed finesse, smoothness, unerring accuracy and forethought at least as much as they needed brute strength and brute courage.

In this case Puhn did well on his own, and when he and pals built more examples. He'd done the complete design on paper before laying a hand on a wrench and he knew where all the best parts came from, so Puhn and Quasar were ready to supply parts and plans for cars using up to two liters of engine, or he'd build a replica for $9,500. That was one quarter of the price of a McLaren and you'd get, oh, half the speed, so it was worth it in that sense but not worth it because you couldn't win any money.

In the event, the Quasar was just about the best little sports racer of its day, while Puhn did better selling parts and writing books on how to make racing cars work.

Moonbeam

Defining a sports car has always been at least as difficult as defining pornography, but in the case of pornography at least one judge said he knew it when he saw it. With sports cars, you not only have to see it, you have to let the rules decide. And you have to have an open mind.

When sports cars became part of the American culture, their acceptance redefined the term. Racing on oval dirt tracks and drag strips and even on the Bonneville Salt Flats was opened to sports cars as those who raced in other places came to like the notion of the two-seat performance car. It came to pass that if the bodywork and equipment conformed, you could run the drags or the lakes just as you could compete on road circuits.

And just as in road racing, the course came to define the horse.

In this case the car began with the enthusiasm of Dean Moon, a lakes racer and early equipment manufacturer. Moon provided aluminum fuel tanks for drag and lake racing, and came up with the spun aluminum wheel disc for streamlining and the trademark Moon Eyes, the stylized eyeball inserts for the "oo" in Moon. And he brought us throttle pedals in the shape of bare feet. *Creative* is the word here.

Moon decided he wanted to go racing in a sports car. He was a hot rodder, at the drags and salt flats, so that's where he went.

OK, so this special wasn't a road racer. Yet it was good at what it did, and it's justified simply because it showed how far and fast one could go with a Devin body.

Moon began with a ladder frame, the classic set of large steel tubes running fore and aft and cross-braced. Front and rear axles were live, located by sets of Watt links and by trailing arms at the front, leading arms at the back, with coil spring-shocks.

The nominal wheelbase was 100 in.; nominal is used here because the wheelbase could be varied by 2 in. in the front and 2 in. in the rear. Weight distribution was fifty-fifty, but it was also nominal because pulling the rear wheels forward and leaving the front wheels alone, or vice versa, altered it as well. The Chevy V-8, in one of several choices of displacement or configuration, went in the middle of the car, with rear drive of course.

Bonneville rules required road-legal lights and such but were fairly free after that and cars were placed on engine size and aspiration, as in supercharged or not.

The two drag racing clubs had something of the same rules, in that they declared a sports car to need two seats and road gear, even to the point of road legality, while at the same time both groups recognized that the serious cars in the class would never be used for anything except that form of racing.

continued on next page

A monster supercharger lived in front of the small-block Chevrolet and was driven off the crankshaft. Moon Eyes were a popular hot rod touch. Road & Track

The Moonbeam was another Devin body, but on a ladder frame with the engine set way back and suspension and tires designed for going straight ahead, at Bonneville, as shown here, or at the drags. Two seats and highway gear like lights made this special a sports racer, by drag and salt lake rules. Road & Track

continued from previous page

Moon ran the Moonbeam at Bonneville and the drags.

He used a selection of engines. They all began as small-block Chevrolet V-8s, but at Bonneville you could hop from class to class as quickly as you could swap engines. The Moonbeam ran at Bonneville with three engines during its career; two were supercharged and each of the three drove the car to better than 200 mph.

Moon fitted a quick-change differential, the Halibrand, and he could switch tires and alter engine tune and shift the car on its wheelbase for traction as opposed to straightline stability. Then it was off to the drags, where the Moonbeam's best was 10.45 seconds for the standing quarter-mile with a trap speed of 145 mph.

The main engine was a stock 4 in. bore with a stroke of 3.75 in. and a displacement of 375 ci. With 8.25:1 compression ratio, Hilborn fuel injection and a GMC 6/71 supercharger (taken from a diesel engine) giving one atmosphere, 14 psi of boost, the engine delivered 640 bhp, at 6000 rpm. At the drags the engine was wound to 9300 rpm, although it wasn't recommended that this be done often.

All this came with a tastefully finished Devin body, with various sections and portions moved around to accommodate the machinery. Moon did immaculate work, and the car was painted a brilliant and lovely yellow, the better to underscore the taillights bracketing the parachute used to slow the car at the drags.

Sport is where you find it.

Shadow

This is a sad and disheartening story, so much so that it must be made plain here that the story is presented as an illustration, a moral fable to instruct those who think success comes easy. It also shows what happened to the art of advancing the technology, once the technology was already advanced.

In August 1969, just about the time the McLarens ruled with their huge alloy engines and with their surplus power, *Road & Track* announced an amazing new Can-Am car. It was the AVS (Advanced Vehicle Systems) Shadow. It came from a California shop run by an informed and articulate man, and it was based on several radical extensions of current thought.

First, tires. We know they'd been getting lower and wider since it was first discovered that the tire surface and the road surface are irregular and gear themselves together, giving better than what was supposed to be perfect traction.

Next, aerodynamics. Wings and spoilers were available, but designers were still trying to arrive at shapes and devices to make the air work and not hold the car back.

So AVS began doubly radical. The team persuaded Firestone, the firm that had begun the tire revolution, to make racing grade tires in 10 in. diameter for the front and 12 in. diameter for the rear. The wheels were two-piece, of aluminum formed on dies and then spun, with the two halves held together by screws and by the studs and nuts that held the wheels to the hubs. The tires were 12 in. wide in front and 17 in. (yes!) wide in back, with overall diameters of 16.5 and 19.4 in. AVS said Firestone promised the tires would be good up to 300 mph.

The tires were that wide to make up for the loss of contact patch that resulted from their being so low. They were so low to reduce the car's frontal area; according to AVS' calculations, the Shadow had frontal area of 13 square feet, against 19 square feet for the then-current McLaren.

There was more. The shape was sleek, with no extra devices and with no intake for the radiator. This last was because the radiators were in the rear, one at each corner, with scoops at the rear corners. The idea was that first, the scoops would bring in cooling air without extra drag and second, under braking with cooling not needed anyway, a vane in the passageway would shut and the scoop would become an air brake. The designer predicted a top speed of 250 mph and said the air brake would be so effective from that speed that the driver wouldn't need to touch the wheel brakes until the car had slowed, presumably instantly, to 225 mph or so. *Road & Track*'s writer paused at that point to say that "we'd best adopt a wait-and-see attitude."

The Shadow's basic structure was safely state of the art, which by 1969 meant a semi-monocoque tub, a collection of torque boxes made of aluminum sheet. The tub was held together by 5,000 rivets, and was anodized black, for strength and looks. The largest boxes were outboard of the engine—that is, toward the back—and held fuel cells of 25 gallons on each side. The front portion, which carried the front body panels, was detachable, on the grounds that it could

The original Shadow hadn't yet run when this photo was taken for a magazine article. The roll bar looks like a nonaerodynamic afterthought, but the low-drag wedge shape and rear corner scoops for the radiator can be seen.
Road & Track

be easily replaced if hit and that it was where most hits were taken.

The engine was conventional: the ubiquitous Chevy V-8. At the time *Road & Track* reported on the Shadow, the AVS team had a selection of alloy Rats, iron Rats and alloy Mice. The iron 427s were for development work and the aluminum engines, which were available only to the right people and for only a short time, were supposed to do the race work proper. The transaxle was an off-the-shelf Hewland five-speed.

The designer planned to keep the suspension stiff and with such limited travel (the limit being forced by the sheer lowness of the car) that the suspension arms were of equal length in front, nearly equal length in back. This would have meant that the wide, flat tire would tip as the car rolled under cornering force, but the magazine was told that cornering at 0.5 g, or one force of gravity, would cause only 0.5 degree of roll so the tires would effectively remain flat on the ground.

Tiny wheels meant tiny brakes. The fronts were Hurst-Airhearts, with the vented rotors turned down to an 8 in. diameter so they'd fit inside the wheels. Because the small rotors would be subject to overheating, there were cooling fans to extract air, and thus heat, from the discs. The rear wheels were larger, but they were enclosed inside the bodywork, so fans were used there, too, with drive from the tailshaft of the gearbox. This provided a faster spin for the fan as the car gained speed.

Back with the basics, the Shadow had an 86 in. wheelbase. No tread or weight was given, but the car was only 23.5 in. high at the highest point, the engine cover (and not counting the ram-style air intake for the engine). The body was a plain wedge, with no attempt at style, and the rake was so extreme because the weight bias was seventy percent on the rear, at rest. This obviously gave traction under power, but with the radiator, engine, gearbox and fuel tanks in the back, the tail would just as obviously wag this dog unless extreme effort was made to use the air to force the rear wheels against the ground at speed.

No fool he, the *Road & Track* reporter wrote that because the Shadow was new and different, the crew would surely have a long and difficult time sorting out the details, and separating the good ideas from the bad ones. Even so, he said, it gives everybody who's interested plenty to think about, and whether or not it wins, it should have influence on coming designs.

It may have had that influence, in sort of a reverse. The Shadow appeared with none other than George Follmer, a good driver and a daring and resourceful one. But owing to tests that showed the body wasn't doing what it was supposed to do, the car had already been fitted with a rear wing. It qualified well off the pace and retired with overheating. Same for the second race of the year. And later the team truck was crashed by a drunk coming the other way ... nobody said luck wouldn't play a part.

In the rest of the world, our old pal Jim Hall and his anonymous crew from Warren, Michigan, unveiled the Chaparral 2J sucker car, with skirts that sealed it to the ground, and with auxiliary motors that vacuumed the air from the car and glued it to the pavement.

Remember when Carroll Shelby got so snippy with his letters to GM? McLaren's Teddy Mayer went

The Shadow's semi-monocoque tub was made of black anodized aluminum sheet in varying thicknesses, riveted together. The main boxes held the fuel cells and bracketed the engine, just visible here behind the light-colored plate at the center rear of the cockpit. Road & Track

The wedge didn't deliver enough downforce to the back of the car, so the team abandoned the low-drag scheme and went with a rear wing, as pioneered by Jim Hall a couple seasons earlier. The car still got the rear corner radiators, though. Road & Track

one better and formally filed a protest, on grounds that the car violated Group 7 rules. Jim Hall simply said he'd explained the whole project to the SCCA before building the car and they'd assured him it was legal. Acting like gents, the SCCA said yes, they'd said what Hall said they'd said and they'd stick to it. On the track, the Chaparral was fastest and broke; the McLarens were strongest and won.

This went on until late in the season, when the SCCA stepped aside and let the FIA declare the 2J illegal on grounds that it had aerodynamic devices above the height restriction. The wings—ol' Jim Hall again—had gone too far. So the McLarens were winning, and now they had virtually no opposition. But wait for the next step; nature abhors a vacuum.

The real story here was that Chevrolet was ready to leave racing, having won everything there was. They were good sports. The top people knew that the 2J had raised the ante, so in a quiet way they graciously stepped back and let the FIA rule so the cost

This was not enough, so the radiators were installed in the wing. Meanwhile the front wheels had grown extensions with turbine blades to ingest cooling air and the rear wheels were out in the airstream as more unique touches were abandoned. Road & Track

and complexity of the racing cars wouldn't get any worse than they were... which was pretty bad. It was a polite and well-executed move.

Tracing the Shadow gets difficult at this point because the public was losing interest in the Can-Am. Surely the prime reason was that the same team, McLaren, rolled over everybody else, while the people like Hall showed sparks of thought but couldn't somehow keep the rig running. Further, the fans had the Trans-Am with those nifty little production-based sedans, and they had Formula A with those stock-block single-seaters. In addition, both series had smaller classes, and it didn't hurt them that the fans drove to the races in Camaros and Firebirds and Alfas and Datsuns, versions of which they could see on the track.

This meant that the magazines didn't report lap by lap and car by car, so the Shadow became invisible. It didn't figure in the reports or the results.

But in May 1971 a former team member, and a disillusioned one at that, told all to *Road & Track*. He said there'd been two cars at first, and AVS began testing early in 1970 and learned then that overheating would be a problem and that the car was skittish. Thus, they added a rear spoiler and, *poof*, up went frontal area by fifty percent and there went the car's initial design advantage.

The spoiler wasn't enough, hence a wing. Then the little tires meant the tallest gears weren't tall enough so the gearbox was reworked, upside down, for an overdrive top. It turned out that the weight was actually seventy-five percent on the back, and curing that required a wing *and* a spoiler. Then the radiators were moved from the rear corners to the wing. But that was protested and eventually the radiator, just one, wound up between the wing and the rear deck. The cooling fans for the brakes kept breaking, resulting in butchery to the bodywork for scoops and hoses, and the parallel suspension was redone to correct for camber during roll, just like the other guys were doing.

Road & Track then said that there had been money problems all along, with creditors guarding the assets and so forth.

The next thing the public knew was that the man behind the AVS Shadow had joined forces with another builder, who'd also challenged the McLarens with a different car. The second project had been more conventional and had come within shooting distance of the rulers. But when the two joined forces, they signed for a new sponsor and the AVS Shadow became the UOP (Universal Oil Products) Shadow. The second venture rolled on Goodyear tires, much taller than those Firestone built as an experiment.

The UOP Shadow did make one bold move, a try at turbocharging the giant Chevy V-8. During testing the monster peaked at 800 bhp. The car made it to the

By the time the car could post a decent qualifying time and give the winners a run, it had normal tires and bodywork and all the things McLaren or Porsche was doing better. Road & Track

races but only after the new backers had spent $160,000 and waited five months.

They probably felt like the guys who invented 6-Up.

While the Shadow was working on a turbocharged Chevy, Roger Penske and his team had made a deal with Porsche, which was looking for a field in which to win. They did a turbocharged Porsche racer, the 917. The team and the factory and the research department spent millions, and the 917 in Can-Am form became the biggest, baddest and meanest racing car in history: Even in 1989, thanks to rule changes and various other limits, the 917 held the record for outrageous excess in pursuit of titles.

In the short run this overwhelmed the new Shadow, the McLarens and everybody else in the Can-Am. In the long run the Can-Am became such an obvious result, even worse than the McLarens before them or the Chaparrals before that, that the series died. And with the Can-Am series went the last venue of the big hairy sports-racing cars.

And with that, we arrive at the end of the special.

This is said with all due respect, and with regret. We began this saga in an era in which the underdog could triumph. We've seen men like Cunningham and Miles and Balchowsky defying convention with their own thinking and their own money. They kept their faith and won against the odds. But by 1972 the aimless blade of science had indeed slashed the pearly gates. Sports car racing—make that the sports car itself—was gone.

Epilogue

This is the era Pete Lyons described as the Nader of Automotive Design.

Exactly why the sport went the way it did, everybody can guess and nobody can say for sure.

One odd possibility surfaced when the McLaren and Porsche teams took reporters for rides. The last of the Can-Am cars took full advantage of turbocharging and aerodynamics. They had 1000 bhp and could corner, brake and accelerate at better than 1 g. From the passenger seat a ride in the Can-Am winner brought to mind Parnelli Jones' description of desert racing: an all-day airplane crash. Riding with Peter Revson or Mark Donohue meant you couldn't hold your head upright through the turns, but you could stand up lying down, feet on the firewall and body above the seat, under braking. It was nearly impossible to believe, let alone describe.

Thing was, from the grandstand the cars were so good and so seemingly in control that they looked calm and deliberate: *thump, thump, thump* and they were around the turn and out of sight.

They were so good in a technical way that they were no fun, they gave no sense of brave men exceeding the limits of fact.

They were boring, to put it bluntly, and because they were boring the series collapsed.

We also have our abdication. We the people gave Congress, or perhaps the hired minions of various agencies supposedly overseen by Congress, veto power over the car. It wasn't buyer and seller anymore. We forced the car companies to give us what we should have wanted, and we drove the small firms out of business because they couldn't afford to meet the rules. We gave ourselves mass transportation, and took away the individualism that created the sports car in the first place. And that meant the end of the sports car in the purist, classic, Porsche Speedster and MG roadster sense.

When support and participation from and by the membership went away, the SCCA and its new rival IMSA threw themselves at the mercy of the car companies. Everybody got to race the same car: a replica built on a space frame, with full race engine but wearing body panels vaguely resembling something for sale to the public. Production car racing became kit car or formula car racing, for professional teams and by professional teams and with the rules arranged so that innovation and creative thinking wouldn't be allowed on the track.

Then there was the re-assignment of talent required by the federal government's new rules. Effort and brains that could have gone into better miles per gallon or braking went into heavier bumpers. And there was insurance. Courts went so far giving money from the insured to those who would take the trouble to claim it that entry fees for races, and even insurance for the street, went beyond belief.

Let me say here that the old cliché about the American public's love affair with the automobile has been wrong for two generations. When everybody had a car, we didn't have an affair: we had a marriage with personal transportation. And in this case, the seventies saw auto design *and* enthusiasm at the lowest ebb since the car was invented. Thus, it may be that the sports car and road racing and homemade racers all went away because not enough people were interested to keep them going.

We may have given ourselves the sport we deserve. It looks as if nobody has built an effective homemade car since 1972 or so, nor won anything major with a homebuilt since 1965. If you did and I missed it, sorry.

What we can hope for is that somewhere, perhaps even now, a couple of bright and energetic kids are getting ready to strip a neat little car, install an engine that's too much for the chassis and invent the road racing special all over again.

Index

2Jr, 15
Abarth automobiles, 196-197
Abarth, Carlo, 197
Aguila, 155-158
Alfa Romeo automobiles, 11
Allard J-2, 18-20
Allard J-2X, 42
Allard JR, 42-43
Allard, Sidney, 18-20, 42-43
Altemus Auto-Banker, 22-24
American Automoible Association (AAA), 23
American Road Race of Champions (ARRC), 191
Ardent Alligator, 11-13
Arkus-Duntov, Zora, 18, 71
Aston Martin automobiles, 79, 96
Auburn automobiles, 8
Austin, Addison, 44
Automobile Racing Club of America (ARCA), 8, 16
Balchoswky, Max, 45, 65, 75, 76-81, 103-105, 118, 128-130, 154, 188
Balchowsky, Ina, 78
Baldwin Mark II, 68-69
Baldwin Special, 14
Baldwin, Bill, 17, 68
Banks, Henry, 142
Barlow, Roger, 25
Barnes, Tom, 51-53, 92, 109, 122, 125, 126, 127, 136, 144, 163, 173
Barneson, John, 75-76
Barneson-Hagemann Chrysler, 75-76
Beach, Gene, 158-160
Beaumont, Charles, 102-103
Beavis Offenhauser, 69
Beavis, George, 69
Beetle, 18
Begra, 158-160
Behra, Jean, 97
Bentzinger, Bob, 49
Bird, Tracy, 51
Bobsy Mark II Ford, 166-169
Bobsy SR-3, 190-192
Bocar XP-4, 105
Bocar XP-5, 100-101
Bocar XP-6, 105-107
Bond, John, R., 5, 89, 130, 142
Bonnier, Jo, 97
Borgeault, 195
Brabham, Jack, 111
Brero, Lou, 73
Broadwell, James, 160
Brock, Pete, 193
Brundage, I. J., 18
Bu-Merc, 10-11
Bugatti automobiles, 8,
Burgess, John, 9
Butterball, 63-64
Butterworth, Archie, 64
Caballo de Hierro, 34-38, 69
Cadillac-Kurtis, 73
California Sports Car Club (CSCC or Cal Club), 16, 96, 117, 168
Canadian-American Challenge Cup (Can-Am), 126
Cannon Special, 14, 22
Carnes, Bob, 100-101, 105-107
Carstens, Tom, 72-74, 119

Chaparral 1, 95, 121, 122-127, 129, 133, 141, 178
Chaparral 2, 171-187
Chaparral 2C, 189-190
Chaparral 2D, 192
Chaparral 2J, 190, 203-205
Chapman, Colin, 50, 140, 171
Chayne, Charles, 10
Cheetah, 130-133
Chinetti, Luigi, 20
Clark, Jim, 184
Cobra, 124
Collier, Barron, Jr., 7, 8
Collier, Miles, 7, 8, 12, 13, 22, 188
Collier, Sam, 7, 8, 22
Comstock-Sadler Mark V, 119-122
Coons, Frank, 153
Cooper automobiles, 55
Cooper Monaco, 123
Cooper, John, 107-108
Cooper-Chevrolet, 144-146
Cooper-Dodge Zerez, 184
Coppel, Al, 22
Corvette Grand Sport, 145, 184
Corvette Special, 154-155
Cozzi Jaguar, 80, 81
Cozzi, Dan, 81
Crawford, Ed, 92
Crawford, Ray, 97
Cro-Sal Special, 184
Crosley, Powel Jr., 28
Cunningham C-2, 24-25
Cunningham C4R, 28-30
Cunningham C4RK, 30
Cunningham C5R, 43-44
Cunningham C6R, 57-63
Cunningham Cooper-Buick, 133-134
Cunningham, Briggs, 5, 10-11, 20-22, 24-25, 28-30, 31, 43-44, 45, 47, 57-63, 74, 89, 117, 133-134, 188

Daigh, Chuck, 51, 79, 80, 92, 93, 97, 108, 109, 111
Dailu, 127-128
Darwin Special, 18
DePalma, Ralph, 10
Deidt, Emil, 18, 92
Devin Super Shillelagh, 97-99
Devin bodies, 53-55, 83, 87, 97, 115, 201
Devin, Bill, 53-55, 98-99
Dietrich, Chuck, 166
Dixon, Freddy, 11
Doane, Dick, 169-170
Dolphin-Abarth, 185, 196-197
Dolphin-Porsche, 160-162
Donner, Frederic G., 180
Donohue, Mark, 195
Drake, Bob, 79
Dryer, Floyd, 28
Durant Special, 191
Durant, Dick, 191, 192

Echidna, 112, 113-117
Edelbrock, Vic, 72
Edmunds, Don "Red," 130-133, 138
Edwards Special, 18
Edwards, Forrest, 66-68
Edwards, Sterling, 18
Eisert, Jerry, 163

Elva automobiles, 118
Ermini automobiles, 53
Evans, Tom, 197
Excalibur, 50-51
Eyerly Crosley, 47-49
Eyerly, Harry, 47-49

Fatima, 9
Federation Internationale de l'Automobile (FIA), 17
Ferrari automobiles, 18, 38, 51, 72, 80, 109
Fitch, John, 24, 47
Flaherty, Jack, 96
Follmer Lotus 23-Porsche, 185
Follmer, George, 184, 185,
Ford GT40, 152
Ford-Cooper (Super Cobra), 152
Fordillac, 20
Formula Continental, 134
Formula Libre, 133
Foyt, A. J., 137, 157, 179, 182
Frame, Fred, 18
Frick, Bill, 20, 128

Gamble, Fred, 95
Genie-BMC, 146-150
Genie-Corvair, 144, 150-152
Genie-Ford, 153, 195
Genie-Olds Mark 8, 152-154
Ginther, Richie, 81, 97, 178
Goossen, Leo, 109
Grady, Henry, 158-160
Graham, Jack, 96
Grant Lotus-Buick, 144
Grant, Jerry, 119, 144, 149
Green, Andy, 173-174
Gregory, Masten, 47, 97
Grierson, Ed, 114-117
Griswold, Frank, 11
Gurney, Dan, 81, 94, 97, 105, 143, 153, 192

Hagemann, Jack, 46, 75, 80, 116
Hall, Jim, 88, 95, 112, 121, 122-127, 141, 143, 171-187, 188, 190, 203-205
Hall, Mike, 185
Hall, Richard, 122
Hamilton, Scott, 166
Hansgen Jaguar, 44-47
Hansgen, Walt, 23, 44-47, 89, 92, 94, 116, 134, 144, 188
Harrison Lotus-Ford, 162-164
Harrison, Doug, 35
Harrison, J. Frank, 163-164
Hauser, Eric, 78, 80, 118, 119
Heuer, Harry, 123, 125
Hilborn, Stu, 87
Hill, Phil, 15, 16-17, 40, 80, 97, 111, 178, 181
Hissom, Ron, 186
Huffaker Chevrolet Special, 87-88
Huffaker, Joe, 146-155
Hulme, Dennis, 190
Huntoon, George, 18
Huntoon-Brundage Special, 17, 18
Hussein, 182
HWM-Chevrolet, 72-74

Ireland, Innes, 144

Jabro Mark III, 160
Jaguar automobiles, 15, 43, 51, 72, 74, 88
Jennings, Bruce, 186

207

Johnson, Sherwood, 38
Jomar, 84, 85-86
Jones and de Camp Crosley, 81-83
Jones, Parnelli, 184

Kangaroo-Fiat, 195, 196
Kellison GT, 101-102
Kellison, Jim, 101-102
Kimberly, Jim, 38
King Cobra, 158
Knoop, Fred, 87-88
Knudsen, Semon E., 180
Kolb, Charley, 95
Krause, Bill, 112, 119
Kurtis 500S, 30-34, 44
Kurtis 500X, 63, 79
Kurtis, Frank, 30-34, 51, 63, 126, 155-158

Ladd, Lem, 9
Lader, Allan, 195
Lagonda-Chrysler, 22
Larson, Bill, 114-117
Le Biplace Torpedo, 27, 28
Le May, Curtis, 39
Le Monstre (Le Manster), 20-22
LeGrand-PBS, 195-196
Lion Cage, 78
Lister, Brian, 88-89
Lister-Chevrolet, 116
Lister-Jaguar, 88-89, 92, 96
Livingston, Duffy, 44
Lola automobiles, 118
Lola-Chevrolet, 195
Lola-Ford, 185
Long Renault Special, 65-66
Long, Al, 65-66
Lotus automobiles, 118
Lovely, Pete, 54, 55-57

MacDonald, Dave, 154-155
Manning Special, 26
Manning, Charles "Chuck", 26
Markleson, Alan, 96
Martin T-1, 113
Martin T-2, 113
Martin T-3, 113
Maserati automobiles, 72, 80, 105, 112
Mayer, Teddy, 203
Mayer, Tim, 142
McAfee, Jack, 51
McKee Chevette, 169-170
McKee, Bob, 169-170
McLaren, Bruce, 133, 144, 189
McLaren-Elva, 185, 189, 192
Mecom, John, 137, 144, 179, 184
Mercedes-Benz automobiles, 10
Mercury Outboard Special, 164, 165-166
Merlin Special, 15
Meyer Special, 28
Meyer, John, 28
MG automobiles, 7, 8
Miles R-1, 39, 40-42
Miles, Ken, 39, 40-42, 47, 57-60, 65, 92, 95-96, 109, 158, 160, 162, 188
Miller Devin-Chevrolet, 96, 97
Miller Devin-Ford, 100
Miller, Ak, 34-37, 53, 55, 69, 80, 96, 97, 100, 101, 106
Miller, Eddie, 136
Millikin, Bill, 64
Mirage, 192-195
Momo, Alfred, 133
Mong, Jerry, 166-169, 190-192
Moon, Dean, 200-202

Moonbeam, 200-202
Morgensen Special, 64-65
Morgensen, Dick, 64-65
Morris Minor Special, 66-68
Moss, Stirling, 111, 118, 119
Muntz, Earl "Mad Man", 32
Murphy, Bill, 63, 79

National Association for Stock Car Racing (NASCAR), 96
National Hot Rod Association (NHRA), 35
Nethercutt, Jack, 192-195
Nissonger KLG Special (Sadler Mark 3), 99-100
Nissonger Special (Sadler Mark 2), 97

Odenburg, Dan, 164, 166
Oehrli, Art, 183
Offenhauser engines, 59-62
Ol' Yaller (Old Yeller) I, 76-81, 103, 104, 118, 119
Ol' Yaller II, 103-105, 112
Ol' Yaller III, 128-130, 133, 143
Ol' Yaller IV, 95, 128-130, 133, 143, 155
Ol' Yaller VII, 143
Ol' Yaller VIII, 130
Ol' Yaller, 126, 130
Old Gray Mere, 8-10
Olson, Warren, 94, 109
OSCA automobiles, 40, 113

Pabst, Augie, 178
PAM Special, 189
Parkinson Jaguar, 26-27, 40
Parkinson, Don, 26-27
Parsons, Chuck, 149, 192
Payne Special, 17-18
Payne, Phil, 17-18
PBX Crosley Special, 49-50
Penske, Roger, 112, 133, 140-146, 172, 178, 179, 188
Peterson, Ted, 126
Pfund, Leonard, 17
Plotkin, Fred, 195
Pollack, Bill, 68-69, 80
Poole, Candy, 49
Pooper, 54, 55-57
Porsche automobiles, 40
Porter Mercedes SLS, 74-75
Porter, Chuck, 74-75
Presley, Elvis, 130
Puhn, Fred, 198-200

Quasar, 198-200
Qvale, Kjell, 158

Rathmann, Jack, 97
Rattenburg Crosley, 82-84
Rattenburg, James, 83-84
Remington, Phil, 18, 109, 111-112
Reuter, John C., 9
Reventlow, Lance, 89-94, 97, 108-112, 125, 134-140, 188
Revson, Peter, 206
Ridenour, Dave, 149, 154, 157
Roberts, Fireball, 179
Rodriguez, Pedro, 143, 153
Rosebud Lotus-Ferrari, 164-165
Ruby, Lloyd, 97, 163
Rutan, Bill, 46
Ruttman, Troy, 34

Saab Special, 197-198
Sadler Chevrolet, 86-87, 95
Sadler Mark 3, 99-100

Sadler, Bill, 86-87, 99, 119-122, 125, 179
Said, Bob, 97
Saidel, Ray, 85-86
Salyer, Ralph, 184, 192
Satcher Special, 17
Satcher, D. G., 17
Scarab Formula One, 105, 107-112
Scarab front-engine, 89-94, 97, 112-113, 141
Scarab-Buick mid-engine, 134-140, 144, 179
SCCA, 13
Schrafft, George, 27
Schroeder, Bob, 63
Seifried Special, 25-26
Seifried, Dick, 25-26
Shadow, 202-205
Sharp, Hap, 121, 126, 127, 143, 172, 175-176, 186, 187
Shelby, Carroll, 80, 81, 94, 97, 112, 124, 158, 180
Shingle, 57-60, 65
Smith, Clay, 33, 37, 63
Spear, Bill, 30, 38, 47
Spencer, Lew, 162
Sports Car Club of America (SCCA), 16, 22-24, 39-40, 96, 117, 132, 168
Staver, John, 114-117
Stevens, Brooks, 50
Stiles, Phil, 27
Stroppe Special, 138-140
Stroppe, Bill, 15, 16-17, 27, 33-34, 37, 44, 51, 63, 138-140, 188
Surtees, John, 179
Sutton, Jack, 74
Swenson, Bob, 196
Swenson, Paul, 196

Tanner Crosley-Saab, 84-85
Tanner, Martin, 84-85
Tappett, Ted, 13
Tatum, Chuck, 46
Teague, Marshall, 96
Team Flat Kat, 93
Telar Special (Zerex Special), 141
Thomas, Bill, 130-133
Thompson, Mickey, 126, 135
Thorne, Joel, 9
Timmons, Paul, 19
Titus, Jerry, 132, 151-152, 169, 188
Tojeiro, John, 87
Travers, Jim, 153
Troutman, Dick, 51-53, 92, 109, 122, 125, 126, 127, 136, 144, 163, 173
Troutman-Barnes Ford Special, 51, 79, 80

United States Auto Club (USAC), 96, 117, 132
United States Road Racing Championship (USRRC), 142
Unser, Bobby, 97

Van Der Feen, Dick, 132
Van Valkenburgh, Paul, 132, 181
Von Neumann, John, 80

Wacker, Fred, 28
Walters, Phil, 13, 24, 28, 30
Ward, Rodger, 133
Webb, Bob, 142
Webster Special, 169, 185

X-Ray Special, 67, 68

Zerex Special, 140-144
Zipper, Otto, 160-162